T0266903

AIRCRAFT AERODYNAMIC DESIGN

Aerospace Series List

AIRCRAFT AERODYNAMIC DESIGN

GEOMETRY AND OPTIMIZATION

András Sóbester and Alexander I J Forrester

Faculty of Engineering and the Environment, University of Southampton, UK

This edition first published 2015
© 2015 John Wiley & Sons, Ltd

Registered office
John Wiley & Sons Ltd, The Atrium, Southern Gate, Chichester, West Sussex, PO19 8SQ, United Kingdom

For details of our global editorial offices, for customer services and for information about how to apply for permission to reuse the copyright material in this book please see our website at www.wiley.com.

Library of Congress Cataloging-in-Publication Data

Sóbester, András.
    Aircraft aerodynamic design: geometry and optimization/András Sóbester, Alexander Forrester.
        pages cm – (Aerospace series)
    Includes bibliographical references and index.
    ISBN 978-0-470-66257-1 (hardback)
    1. Airframes.   2. Aerodynamics.   I. Forrester, Alexander I J.   II. Title.
    TL671.6.S58 2014
    629.134′1–dc23
                                                        2014026821

A catalogue record for this book is available from the British Library.

ISBN: 9780470662571

Set in 10/12pt Times by Aptara Inc., New Delhi, India

1   2015

Contents

Series Preface

The field of aerospace is multi-disciplinary and wide ranging, covering a large variety of products, disciplines and domains, not merely in engineering but in many related supporting activities. These combine to enable the aerospace industry to produce exciting and technologically advanced vehicles. The wealth of knowledge and experience that has been gained by expert practitioners in the various aerospace fields needs to be passed onto others working in the industry, including those just entering from university.

The *Aerospace Series* aims to be a practical, topical and relevant series of books aimed at people working in the aerospace industry, including engineering professionals and operators, allied professions such commercial and legal executives, and also engineers in academia. The range of topics is intended to be wide ranging, covering design and development, manufacture, operation and support of aircraft, as well as topics such as infrastructure operations and developments in research and technology.

Aerodynamics is the fundamental science that underpins the world-wide aerospace industry and enables much of the design and development of today's highly efficient aircraft. Much effort is devoted to the design of new aircraft in order to determine the wing and tail surface geometry that gives the optimum aerodynamic performance.

This book, *Aircraft Aerodynamic Design: Geometry and Optimization*, covers a range of different aspects of geometry parameterization that is relevant for aircraft lifting surface design. The emphasis is on the efficient construction of an aircraft geometry which can then be coupled to any flow solver and optimization package; however, most of the concepts can be applied to any engineering product. Starting with the underlying principles of geometric parameterization, the reader is taken through the fundamentals of 2D aerofoil optimization onto 3D wing synthesis and the computation of design sensitivities. The most important concepts are illustrated using a basic aerofoil analysis and a human powered wing design. All of the key ideas throughout the book are demonstrated using computer codes, making it easy for the readers to develop their own applications. The book provides a welcome addition to the Wiley Aerospace Series and complements other books on aerodynamic modelling and conceptual aircraft design.

Peter Belobaba, Jonathan Cooper and Allan Seabridge

Preface

In July 1978 the *Journal of Aircraft* published a paper titled 'Wing design by numerical optimization'. The authors, Raymond Hicks of the NASA Ames Research Center and Preston Henne of the Douglas Aircraft Company, had identified a set of functions with 'aerofoil-like' shapes, which, when added to a baseline aerofoil in various linear combinations, generated other 'sensible' aerofoil shapes.

This, as a principle, was not new. After all, the National Advisory Committee for Aeronautics was already experimenting with *parametric aerofoils* in the 1930s. The formulation described by Hicks and Henne (1978) was a new aerofoil family generated in a novel way – building an aerofoil out of weighted shapes, much like one might build a musical sound from multiple harmonics. But this was not the real novelty; how they proceeded to use it was.

Combining the incipient technology of numerical flow simulation (they used a two-dimensional model) with a simple optimization heuristic and their new parametric geometry they performed an automated computational search for a *better aerofoil shape*.

Here is the idea that thus began to take shape and commence its ascent along the technology readiness level (TRL) ladder of the aerospace industry. A parametric geometry is placed at the heart of the aircraft design process. The *design variables* influencing its shape are adjusted in some systematic, iterative way, as dictated by an optimization algorithm. The latter is guided by a design performance metric, resulting from a physics-based simulation run on an instance of the parametric geometry.

The TRL rise was to be a slow one, for two reasons. First, because in a world largely reliant on drawing boards for years to come, this was a disruptive idea that would encounter much resistance in this notoriously risk-averse industry. Second, none of the links in the chain of tools required (numerical flow analysis, computational geometry and efficient optimization techniques) would be really ready for some fast optimization action until well into the 1990s.

There is a maxim known by most practitioners of the art, which states that an optimization algorithm will find the slightest flaws in the analysis code (usually comprising a mesher and a partial differential equation solver) and in the geometry model; that is, it will steer the design process precisely towards their weak areas.

This is not (only) due to Sod's law – more fundamental effects are at play. Most computational analyses have a domain of 'safe' operation, outside of which they will either predict unphysically good or unphysically bad performance. Straying into the latter type of area will thus be a self-limiting deviation, but the former will lure the optimizer into 'discovering' amazingly good solutions that do not actually exist in 'real' physics. Sometimes these are obvious (what rookie optimization practitioner has not 'discovered' aerofoils that generate thrust instead of drag?), but more subtle pitfalls abound, and highlighting these remains a challenge in the path of the ubiquitous use of this technology.

Along similar lines, parametric geometry modelling has its own pitfalls, deceptions and hurdles in the path of effective optimal design, and how to avoid (at least some of) them is the subject of this book.

Some of the principles discussed over the pages that follow can be applied to the geometry of any engineering product, but we focus on those aspects of geometry parameterization that are specific to external aircraft surfaces wetted by airflow. Some of the ideas are therefore linked to aerodynamics, and so we will touch upon the relevant aspects of aircraft aerodynamic design – from an engineering perspective. However, *this is not a book on aircraft aerodynamics*, and, for that matter, nor will it provide the reader with a recipe on how to design an aeroplane. Instead, it is an exposition of concepts necessary for the construction of aircraft geometry that can exploit the capabilities of an optimization algorithm.

The reader may wish to peruse the text simply to gain a theoretical appreciation of some of the issues of aircraft geometry parameterization, but there is plenty to get started with for the more practically minded too. All key concepts are illustrated with code, which can be run 'as is' or can form a building block in the reader's own code. After lengthy deliberations we selected two software platforms to use for this: Mathworks MATLAB® and Python. Some of the Python code calls methods from the OpenNURBS framework, which can be accessed through *Robert McNeel & Associates Rhino*, a powerful, yet easy to use, lightweight CAD package. Some of the code is reproduced in the text to help illustrate some of the formulations – in each case we selected one of the platforms mentioned above, but in most cases implementations in the others are available too on the website [www.wiley.com/go/sobester] accompanying the book.

Here is a brief sketch of the structure of this book.

After discussion of the general context of aircraft shape description and parameterization (*Prologue*), in the following chapter (*Geometry Parameterization: Philosophy and Practice*) we discuss the place of parametric geometries in aircraft design in general and we start the main threads that will be running through this book: the guiding principles of parametric geometry construction and their impact on the effectiveness of the optimization processes we might build upon them.

We next tackle the fundamental building blocks of all aircraft geometries, first in two dimensions (the chapter titled *Curves*), then in three (*Surfaces*). Two-dimensional sections through wings (and other lifting surfaces) are perhaps the most widely known and widely discussed aerodynamic geometry primitive, and we dedicate three chapters to them: a general introduction (*Aerofoil Engineering: Fundamentals*), a review of some of the key *Families of Legacy Aerofoils* and, arriving at the concept at the heart of this book, *Aerofoil Parameterization*.

Another classic two-dimensional view of aerodynamics is tackled in the chapter titled *Planform Parameterization*, thus completing the discussion of all the primitives needed to build a three-dimensional wing geometry – which we do in the chapter *Three-Dimensional Wing Synthesis*.

The ultimate point of geometry parameterization is, of course, the optimization of objective functions that measure the performance of the object represented by the geometry. Recent years have seen a strong push towards making this process as efficient as possible, and one of the enablers is the efficient computation of the sensitivities of the objective function with respect to the design variables controlling the shape. A number of ways of achieving this are discussed in the chapter titled *Design Sensitivities*.

The most important concepts are illustrated via examples throughout the book, but there are two more substantial such examples, which warrant chapters of their own: *Basic Aerofoil Analysis: A Worked Example* and *Human-Powered Aircraft Wing Design: A Case Study in Aerodynamic Shape Optimization*.

We then bring matters to a close by looking ahead and discussing the area where geometry parameterization is most acutely in need of further development – this is the chapter titled *Epilogue: Challenging Topological Prejudice*.

Parametric geometry is a vast subject, and a book dedicated even to one of its subsets – in this case, the parametric geometry of the external shape of fixed-wing aircraft – is unlikely to be comprehensive. We hope that, beyond a discussion of the formulations we felt to be the most important, this book succeeds in setting out the key principles that will enable the reader to 'discover', critically evaluate and deploy other formulations not discussed here. Moreover, it should assist in creating new models – essential building blocks of the design tools of the future.

Finally, we would like to acknowledge some of those who helped shape this text through discussions and reviews: Jennifer Forrester, Brenda Kulfan, Andy Keane, Christopher Paulson, James Scanlan, Nigel Taylor, David Toal and Sebastian Walter. We are also indebted to Tom Carter and Eric Willner at Wiley, whose patience and support made the long years of writing this book considerably easier.

Disclaimer: The design methods and examples given in this book and associated software are intended for guidance only and have not been developed to meet any specific design requirements. It remains the responsibility of the designer to independently validate designs arrived at as a result of using this book and associated software. To the fullest extent permitted by applicable law John Wiley & Sons, Ltd. and the authors (i) provide the information in this book and associated software without express or implied warranties that the information is accurate, error free or reliable; (ii) make no and expressly disclaim all warranties as to merchantability, satisfactory quality or fitness for any particular purpose; and accept no responsibility or liability for any loss or damage occasioned to any person or property including loss of income; loss of business profits or contracts; business interruption; loss of the use of money or anticipated savings; loss of information; loss of opportunity, goodwill or reputation; loss of, damage to or corruption of data; or any indirect or consequential loss or damage of any kind howsoever arising, through using the material, instructions, methods or ideas contained herein or acting or refraining from acting as a result of such use.

András Sóbester and Alexander I J Forrester
Southampton, UK, 2014

1

Prologue

Geometry is the *lingua franca* of engineering. Any conversation around a nontrivial design problem usually has even the most articulate engineer overcome, within minutes, by the desire to draw, sketch or doodle. Over the centuries the sketching tools have changed. However, Leonardo da Vinci wielded his chalk and pen for the same reason why today's engineers slide their fingertips along tablet computer screens, deftly creating three-dimensional geometrical models and navigating around them: the functionality and performance of an engineering product depends, to a very large extent, on its shape and size; that is, on its *geometry*.

Different fields of engineering place different levels of emphasis on geometry, but perhaps none focuses on it more sharply than the aerodynamic design of aircraft. The goal of the aerodynamics engineer is to create an object that, when immersed in airflow, will change the patterns of the latter in a desirable fashion,[1] and this is most readily achieved through the shaping and sizing of the object.

Sadly for aerodynamic engineers, their freedom to play with the form of the aircraft's flow-wetted surfaces is often curtailed by other departments competing for influence over the same piece of real estate: structural engineering, cost modelling, propulsion, control systems, cabin and payload, and so on. Moreover, the aerodynamic performance of an aircraft is usually multifaceted too: different phases of the same mission tend to drive the external shape in different directions. This tension between competing objectives is usually resolved in one of three ways:

1. *One of the goals trumps all others.* The shape usually gives this away – it is immediately clear to the trained observer that one interest drove the design of the aircraft and the others had to operate within very strict constraints defined by it.

   Consider Figure 1.1 as an example. It shows three unmanned aircraft. One has a delicate-looking, sleek airframe with very long and narrow wings: a glider (sailplane), the design of which was driven by the single-minded desire to maximize endurance. The structural and

---

[1] Aerodynamics engineers are also unique amongst the general public in regarding an aircraft as a stationary object, with the atmosphere 'flying' past it.

Aircraft Aerodynamic Design: Geometry and Optimization, First Edition. András Sóbester and Alexander I J Forrester.
© 2015 John Wiley & Sons, Ltd. Published 2015 by John Wiley & Sons, Ltd.

Figure 1.1 Three unmanned air vehicles (UAVs), three main design drivers: NASA Towed Glider Air-Launch Concept (left, NASA image) – long endurance at low speeds; AeroVironment RQ-11 Raven (top right, USAF image) – structural robustness to cope with battlefield conditions; Boeing X-51 (USAF image) – extremely high speed.

cost engineers would no doubt have liked to have seen shorter, stubbier wings, upon which the air loads generate lighter bending moments, but this was a case of shape design in the service of aerodynamic efficiency, with little more than a glance toward other objectives.

The top right image shows a soldier launching a low-altitude, low-endurance surveillance platform. The 'boxy' fuselage and the short, wide wings identify this as a design driven by a desire for ruggedness and enough spare structural strength to allow the aircraft to cope with the rough handling likely in a battlefield environment – at the expense of aerodynamic efficiency.

Finally, the third picture shows a hypersonic research aircraft. It is not able to carry any payload, it has no landing gear (it is a single-use vehicle) and its endurance is measured in seconds. But it does 'ace' one objective: speed. Every feature of its geometry says 'designed for a hypersonic dash' (more on this extraordinary vehicle in the next chapter).

2. *A compromise* results, which balances all the competing goals. How to analyse all the trade-offs involved and how to make design decisions based on them is the core question of modern engineering design and we will discuss some of the relevant techniques in Chapter 2.

3. *The 'all things to all departments' solution*. The aircraft, or aircraft subsystem, is actually several designs rolled into the same packaging, with each design optimized for a particular goal. An in-flight 'morphing' process mutates the geometry from one shape to another, depending on the phase of the mission. Perhaps the most common embodiment of this principle is the high-lift system that enables many aircraft to cruise efficiently at high speed, but also generates sufficient lift at the low speeds typical of take-off and landing (see Figure 1.2).

Figure 1.2 Three wing geometries packaged into one and able to morph from one to another: the trailing edge of the wing of a Boeing 787-8 airliner in (from left to right) cruising flight, final approach and touch-down (photographs by A. Sóbester).

This is a complex problem for the designer, as the challenge is not only to design multiple geometries, each optimized for, say, different flow regimes, but also to choreograph the transition process – all this without exceeding weight, cost and complexity constraints.

Ultimately, the external geometry of an aircraft results from the systematic analysis of the physics behind all the relevant objectives and constraints and the application of engineering optimization techniques to the resulting data to generate a solution that satisfies the design brief. There is an almost endless diversity of design briefs, of different relative importances of the objectives, of physics-based performance simulation capabilities and of design techniques, the result being an immense range of shapes – see Figure 1.3 for a small selection. This diversity is a measure of how serious a mathematical challenge shape modelling is.

In terms of an aeronautical engineer's relationship with geometry, much has changed since da Vinci began sketching flying vehicles. This technological transformation has been a nonlinear one too, with the migration from a two-dimensional drawing board into the three-dimensional space of virtual geometries stored in the memory of a computer constituting the greatest leap. However, the most important aspect of this revolution that began to gather momentum in the 1970s and 1980s was *not* the extension into the third dimension.

Rather, with the advent of the first computational geometry engines, engineers were suddenly able to make local, as well as large-scale, *changes* to the emerging geometry in a systematic and time-efficient manner. This step was particularly significant at the level of whole aircraft geometries – no longer did the change of a simple dimension on one component mean that tens or hundreds of other blueprints had to be redrawn. Instead, the computational geometry cascaded the changes as appropriate, and thus new configurations could be generated in a matter of seconds.

At the component level, generating hundreds or thousands of different candidate shapes became possible, which, in combination with the emerging field of numerical analysis codes capable of simulating the performance of these shapes, gave engineers a powerful design tool.

This new capability created a need for developments in the mathematics of shapes of relevance to aerodynamic design. A new breed of curve and surface models was needed that was equipped with simple handles: *design variables*, the values of which could be altered, leading to intuitive shape changes. This had previously been of little interest, as redrawing a

Figure 1.3 External (or *outer mould line*) geometry – even within the realm of fixed-wing aircraft, the variety and sophistication of surface shapes poses a serious geometry modelling and optimal design challenge (photographs by A. Sóbester).

blueprint or sanding the wind tunnel model of a geometry to a slightly adjusted shape had not required such formalisms.

The Hicks–Henne basis functions mentioned in the Preface (we will return to it later in this text) were one example of such developments. At the same time a much older idea gained a 'second life'.

In the 1930s, long before the earliest electronic computers, the National Advisory Committee for Aeronautics (NACA) had developed a wing section model defined through two pairs of parametric polynomials. By altering the parameter values, a family of designs could be built up, comprising airfoils suitable for a wide range of applications. As we shall see in Chapter 6, members of this family are still 'in service' eight decades later on a number of aircraft. But, perhaps even more importantly, in the new era of computational geometry modelling and numerical flow simulation the NACA sections suddenly became the template for just the type of mathematical formulation demanded by the new technology. The three components of the modern computational design process architecture, as we know it today, crystallized: a parametric shape description, a physics-based simulation code capable of measuring the performance of a given instance of the parametric shape and a systematic means of varying the parameters in a way that enabled progress in an efficient manner towards the optimum solution.

This book is about the intellectual descendants of the NACA wing section model: parametric geometries that make the most of the combined capabilities of analysis codes and optimization heuristics. In addition to reviewing existing formulations and their place in the aerospace engineer's tool set, we also set out the principles we recommend to be followed in designing new formulations. And, in an age when the computational cost of performance analysis is falling as predicted by Moore's law and optimization heuristics represent a fully mature technology, the importance of the effective use of the third element, geometry parameterization, cannot be overstated.

2

Geometry Parameterization: Philosophy and Practice

The subject of this book is the parametric geometry of bodies that are of interest in aerodynamic design. We use the term 'geometry' to cover two aspects of the mathematical representation of such bodies: their shape and scale (or size). Before we begin our exploration of the tremendous variety of aerodynamic shapes, we endeavour to clarify the place of both concepts in aerodynamics.

2.1   A Sense of Scale

2.1.1   Separating Shape and Scale

It is perhaps a little counter-intuitive to begin a chapter (and, indeed, a book) on geometry parameterization with a brief foray into fundamental aerodynamics, but the idea of separating the two fundamental building blocks of geometry is grounded in aerodynamics. We shall thus set out the reasons for this in what follows, in the context of the basic mechanics of viscous, incompressible flow.

At this point, then, the reader anxious to dive straight into the matters at the heart of this text – that is, how to build geometry models suitable for efficient optimization – may wish to skip the remainder of this section, simply taking note of the fact that it is often convenient, from a design point of view, to create some degree of isolation between the shape and the scale of a geometry. Those wishing to know *why*, should read on.

Consider the body immersed in viscous, incompressible flow, as shown in Figure 2.1. The velocity field within the fluid is described by the momentum equation

$$\frac{\partial \Omega}{\partial t} + \nabla \times (\Omega \times v) = \frac{\mu}{\rho} \nabla^2 \Omega, \tag{2.1}$$

Aircraft Aerodynamic Design: Geometry and Optimization, First Edition. András Sóbester and Alexander I J Forrester.
© 2015 John Wiley & Sons, Ltd. Published 2015 by John Wiley & Sons, Ltd.

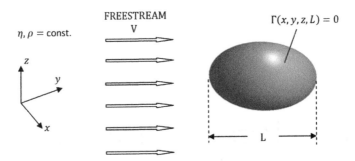

Figure 2.1 A thought experiment: incompressible flow around a body with some characteristic dimension L.

where the vorticity $\Omega = \nabla \times v$.[1] As the fluid is considered incompressible, we assume that the divergence of the velocity field is zero ($\nabla \cdot v = 0$, $\rho = \text{const.}$) – this is true when the Mach number is small ($M \ll 1$).

In terms of the boundary conditions of (2.1), we assume that at a long distance away from the body the velocity of the flow (say, parallel with the x-axis) is V and that it vanishes on the surface of the body. The shape of the body is described in terms of the Cartesian coordinates of the space as $\Gamma(x, y, z, L) = 0$, where L is some characteristic dimension.[2] Thus, we set

$$v_x(x, y, z) = v_y(x, y, z) = v_z(x, y, z) = 0, \forall(x, y, z) \text{ satisfying } \Gamma = 0. \qquad (2.2)$$

Let us now consider a new system of units to describe this flow field. If we express all lengths in terms of L and all velocities in terms of the freestream velocity V (and, implicitly, time in units of L/V), we can make the following substitutions:

$$x \to x'L, \quad y \to y'L, \quad z \to z'L, \qquad (2.3)$$

as well as

$$v \to v'V \quad (v' = 1 \text{ in the freestream}) \qquad (2.4)$$

and

$$t \to t'\frac{L}{V} \qquad (2.5)$$

(we use the same notation for all other quantities; that is, the prime denotes a quantity expressed in the new set of units).

---

[1] For a complete derivation and a more detailed discussion of the line of thought set out below, see the legendary notes by Feynman *et al.* (1964).

[2] For example, a sphere of diameter L could be described in this way as $\Gamma : x^2 + y^2 + z^2 - L^2/4 = 0$.

In order to transform (2.1) to the new units, we look at vorticity first:

$$\Omega = \nabla \times v = \begin{vmatrix} \mathbf{i} & \mathbf{j} & \mathbf{k} \\ \frac{\partial}{\partial x} & \frac{\partial}{\partial y} & \frac{\partial}{\partial z} \\ v_x & v_y & v_z \end{vmatrix} = \begin{vmatrix} \mathbf{i} & \mathbf{j} & \mathbf{k} \\ \frac{1}{L}\frac{\partial}{\partial x'} & \frac{1}{L}\frac{\partial}{\partial y'} & \frac{1}{L}\frac{\partial}{\partial z'} \\ v_x'V & v_y'V & v_z'V \end{vmatrix}. \tag{2.6}$$

This is a neat way of writing the curl of the velocity field, though it is a slight abuse of the determinant notation. Nonetheless, the manoeuvre is not a particularly dangerous one, so long as we are aware that the 'products' in the development of the determinant actually mean 'applications of the differential operators to the terms in the third row':

$$\Omega = \mathbf{i}\left[\frac{V}{L}\frac{\partial v_z'}{\partial y'} - \frac{V}{L}\frac{\partial v_y'}{\partial z'}\right] - \mathbf{j}\left[\frac{V}{L}\frac{\partial v_z'}{\partial x'} - \frac{V}{L}\frac{\partial v_x'}{\partial z'}\right] + \mathbf{k}\left[\frac{V}{L}\frac{\partial v_y'}{\partial x'} - \frac{V}{L}\frac{\partial v_x'}{\partial y'}\right]. \tag{2.7}$$

Thus, the vorticity expressed in terms of the new units of velocity and length can be written as

$$\Omega = \frac{V}{L}\begin{vmatrix} \mathbf{i} & \mathbf{j} & \mathbf{k} \\ \frac{\partial}{\partial x'} & \frac{\partial}{\partial y'} & \frac{\partial}{\partial z'} \\ v_x' & v_y' & v_z' \end{vmatrix} = \frac{V}{L}\nabla' \times v' = \frac{V}{L}\Omega'. \tag{2.8}$$

Applying the substitutions to (2.1) now, we arrive at

$$\frac{\partial \Omega'}{\partial t'} + \nabla' \times (\Omega' \times v') = \frac{\mu}{\rho VL}\nabla^2\Omega'. \tag{2.9}$$

As a result of this normalization of the problem we now have an equation that *isolates the effect of the scale* (size) of the body on the velocity field. The scale is encapsulated in the coefficient of the right-hand-side term. This coefficient, which uniquely holds the scale information (or, rather, the inverse of this coefficient), is, of course, the *Reynolds number*:

$$Re = \frac{\rho VL}{\mu}. \tag{2.10}$$

2.1.2   Nondimensional Coefficients

This identification of the effect of scale on the flow field is especially important in experimental aerodynamics. The Reynolds number tells us how we can compensate for the difference in scale between real aircraft and their wind tunnel models: the model-scale flow field will be similar to the real flow field if we increase the flow velocity, increase the density (typically by a reduction of the temperature in *cryogenic* wind tunnels) or (less practically) decrease the viscosity. All this, of course, is only true if the Mach number stays below about 0.6 – we must not forget that the compressibility term has been omitted from (2.1) and at higher Mach numbers this would invalidate the analysis (ρ will no longer be a constant).

Figure 2.2   Kármán vortex streets can appear at much larger length scales. Shown here is a pattern of stratocumulus clouds around the Pacific island of Guadalupe [NASA image].

These observations are based on a purely mathematical reasoning, but the Reynolds number has a more germane physical interpretation. It is the ratio of the inertial and viscous forces in the flow (the inertial forces are proportional to the density and the square of the velocity, while the viscous forces are proportional to the dynamic viscosity and the velocity and inversely proportional to the length scale); as such, it offers an insight into the influence of scale on the flow field from a different perspective: that of the turbulence within the flow field in the vicinity of the body.

Let us examine this through a (geometrically) simple example, that of a set of long cylinders of a range of diameters D, exposed to airflow perpendicular to their axes of symmetry. The conditions correspond to sea level according to the International Standard Atmosphere model and the speed of the flow is 1 m/s (slow walking pace). Let us consider the major qualitative differences between the airflows around cylinders of various diameters:

- $D = 0.3$ mm ($Re \simeq 20$)   The streamlines flow smoothly around and rejoin behind the cylinder, but a pair of vortices sits just behind it.
- $D = 1$ mm ($Re \simeq 70$)   One of the vortices separates from the back of the cylinder and travels downstream with the airflow. In its place a new vortex forms – meanwhile, another vortex breaks off on the other side. The result of this constant production of new vortices on alternating sides is a stream of vortices travelling downstream with the flow, known as a *Kármán vortex street* (see Figure 2.2). In any given point the flow velocity varies with time, but it does so in a periodic fashion.
- $D = 5$ mm ($Re \simeq 340$)   An increasingly long region downstream of the cylinder is now turbulent – the variation of the velocity in any given point is now chaotic, though there still is an underlying periodic component (it is, however, corrupted by 'random' noise).
- $D = 100$ mm ($Re \simeq 680$)   The boundary layer around the cylinder begins to become turbulent and, as we move to increasingly larger cylinders, the turbulent portion of the boundary layer stretches further and further upstream.
- $D = 10$ m ($Re \simeq 7 \times 10^5$)   The boundary layer has now become fully turbulent; so is the wake, the flow there is now chaotic and the eddies are three-dimensional (3D).

The above classic experimental result focuses on the effect of scale from a purely qualitative point of view. Much of it is also of a somewhat academic interest, as the numbers indicate that the majority of flows of any importance in aircraft design are likely to be in the range of Reynolds numbers associated with fully turbulent boundary layers. It is also clear that from an aerodynamic design perspective we need a more quantitative approach to isolating shape from scale.

To this end, the aerodynamic characterization of aircraft relies heavily on nondimensional derivatives, such as drag coefficients, lift coefficients and pitching moment coefficients. These replace forces and moments related to a geometry, which depend on its length scales, with dimensionless numbers that do not. These allow the characterization of a 'back of a napkin' dimensionless sketch of a geometry even before we know the exact dimensions and, more importantly, allow comparisons between different geometries at the early stages of a design process.

There are, of course, limits to the validity of such comparisons. While, for instance, the drag coefficient C_D neatly moves the scale effects out of sight, we still need to be mindful of the sorts of large, qualitative changes in the physics of the flow that we saw in the case of the cylinders. In other words, C_D itself varies with Reynolds number and, as compressibility effects kick in, varies with Mach number. For a particular design point, where our candidate geometries will experience very similar flow conditions and very significant changes in size are unlikely, C_D is still a useful objective function in an optimization context.

Of course, there are cases when we wish to compare, say, aircraft featuring different geometries and flying at significantly different design points. For instance, we may have a specific target range, but we might be considering a range of speeds from, say, low transonic to low supersonic. In this case an explicitly Mach-number-dependent aerodynamic efficiency metric may be of use: ML/D, the product of the Mach number and the lift-to-drag ratio, which the optimization process would seek to maximize.

Thus, we have a basic set of aerodynamic measures of merit, which we can use to characterize a geometry. A geometry we can characterize is a geometry we can also endeavour to improve by optimizing these metrics; and the key feature of a geometry model that will enable such an optimization exercise is that it has to be *parametric* – this concept lies at the heart of this book, and it is the subject of the next section.

2.2 Parametric Geometries

When presented with a design problem and a pen and a blank sheet of paper, few engineers can resist drawing a tentative, conceptual sketch. Aerodynamic design engineers are no exception. Nevertheless, it is not entirely impossible to imagine an aerodynamic design process that begins with a few 'back of the envelope' calculations that do not necessarily involve a visual representation of the artefact. One could speculate that in the majority of cases this is only possible because some form of broad-brush understanding of the shape and the topology of the object that is to result from the design process exists in the mind(s) involved, but that is an argument best left to more philosophical treatises.

Here, we shall be content to observe that, with the possible exception of some initial conceptual studies, every aerodynamic design process is centred around a geometry; that is, around a set of points, curves and/or surfaces. In other words, starting with at least the *preliminary*, if not the *conceptual* phase of a design process, it is the geometry that is actually being designed, or it is at least a significant aspect of what is being designed.

If a design process is to carry the tag 'optimal' – and most modern aerodynamic design processes do, if not always with great sophistication or even under this explicit label, but at least as a matter of principle – then this central geometry also needs to be *parametric*.

Let us make it clear from the outset that this is a word laden with many different meanings, and misunderstandings can be rather confusing. For instance, consider the following plane curve:

$$\gamma = (x, y) = (\cos u, \sin u), \quad u \in [0, 2\pi]. \tag{2.11}$$

From a mathematician's perspective, this is the parametric curve (in terms of the parameter u) of a circle of radius one. From an engineer's perspective – and this is the usage we adopt here – it is not a parametric geometry, because there are no knobs to twiddle, no obvious handles for changing the shape or size of the circle. Here is an engineer's parametric circle:

$$\gamma(r) = (x, y) = (r \cos u, r \sin u), \quad u \in [0, 2\pi], \quad r \in [r_{min}, r_{max}]. \tag{2.12}$$

This is a circle of radius r; that is, a geometry that changes in response to changes in the *geometrical design variable* or *geometrical design parameter* r. $\gamma(r)$ is, in engineering terms, a *parametric geometry*, and it would still be referred to as one if u was dropped through some rearranging leading to, say, the form

$$\gamma(r) : x^2 + y^2 = r^2. \tag{2.13}$$

To be completely precise, the latter is an *implicit parametric geometry*, as the ordinate y values can only be obtained by solving an equation. These can sometimes be converted into *explicit parametric geometries* – in this particular case this is doable by breaking it down into two semicircles, yielding $y = \pm\sqrt{r^2 - x^2}$. In all of these cases, any r from within its specified range (the range forms an integral part of the parametric geometry) can be inserted into the equation to generate a specific geometry, fixed in shape and size – this operation is known as *instantiation* and, for example, $\gamma(2.5)$ will be referred to as an *instance* of $\gamma(r)$.

Given some metric $f(\gamma) = f[\gamma(r)] = f(r)$ related to the circle (for instance, the drag of a fuselage of that cross-section divided by the amount of payload it could envelop), an optimization problem could be formulated, where we seek to minimize f with respect to the design variable r, resulting in the minimum value of f at some optimum radius r_{opt}:

$$f(r_{opt}) = \min_{r \in [r_{min}, r_{max}]} f(r). \tag{2.14}$$

The optimization process is thus some systematic way in which the variable values are iteratively altered to find their value (and thus the shape) that minimizes or maximizes the objective or goal function.[3] Incidentally, *the design variable is never 'optimized', the objective function*

---

[3] There is an alternative school of thought, which regards γ as a free curve (or surface) and f some *functional*, which measures γ according to some objective metric. The optimum γ is then sought without making explicit algebraic assumptions about its form. This calculus of variations view of optimization lies at the heart of *adjoint-based optimization*, which we shall return to later in Section 10.2.2.

is. It is therefore not correct to refer to the 'optimization of (the variables defining) an aircraft for endurance' – the correct phraseology is 'optimization of the endurance of the aircraft'. In some cases the identities of the objective and the variables may be obvious, in others, painstakingly correct terminology may be the only way of avoiding misunderstandings. In that spirit, here is some more inevitable punctiliousness in the shape of a checklist the designer should apply to a chosen set of design variables before unleashing an optimization algorithm on the resulting problem.

2.2.1   Pre-Optimization Checks

Item 1: Are All Chosen Parameters Genuine Design Variables?

It is worth considering briefly the place of geometrical design variables, which form the central subject of this discussion, within the broader family of *design variables* in general. One could take the somewhat uncompromising view that a design variable is either geometrical or it defines a material choice – our main concern here being the design of aerodynamic surfaces, this is actually not an unreasonable stance to take. A concession could be made to include catalogue-number-type variables (for instance, the catalogue number of a rivet could be a design variable), though it could be argued that those are just conveniently concise packages encapsulating several geometrical variables, so they actually belong to the latter category.

More importantly, though, it is important to separate the concept of a design variable from that of various metrics associated with a design, which – and herein lies some scope for terminological confusion – also vary during the design process. For instance, specific fuel consumption might be a key parameter in gas turbine engine design, but it is not a design variable. Nor is bypass ratio a design variable in the sense of the geometry-centred design processes we are dealing with here, though – scope for more confusion – the first, conceptual, non-geometric phase of the design of a turbofan engine might well view bypass ratio as a design variable.

A simple test to apply to a set of parameters related to a design, which might determine whether they are higher level metrics or design variables (or a mixture), is to consider what would happen if several engineers were issued with these sets of parameters and their unambiguous definitions and they were told to come up, working independently, with the corresponding design blueprints. If you can only imagine them returning with identical blueprints, then you have a set of genuine design variables (of course, this is all based on the assumption that these hypothetical engineers do not make mistakes). Conversely, the possibility of dissimilar designs emerging at the end of this thought experiment would be a foolproof sign that one or more of the parameters are higher level metrics and not design variables.

Item 2: Does the Parameterization Make Mathematical Sense?

More specifically, one might call this an *under- and overparameterization check*. It is a type of thought experiment that may serve as another health test that must be applied to a putative set of design variables.

An underparameterized geometry has fewer variables than needed for its unambiguous construction, and an overparameterized geometry has too many; that is, there are two or more conflicting numbers in the parameter list. This is not about personal preferences or judgement regarding the impact on the optimization problem of removing or adding variables – we shall discuss that later. Instead, this is about the mathematical validity of the parameterization – a right or wrong issue.

For example, there are several right ways of parameterizing a triangle – two angles and the length of the side they share, three side lengths, and so on – and several wrong ones too.

For instance, a triangle defined by two side lengths and two angles will be overparameterized because a three-member subset of this parameter vector will define the triangle unequivocally, so the fourth one is at best unnecessary (if it happens to fit in with the first three) and at worst is in conflict with the triangle defined by the other three. Conversely, specifying, say, only its three angles would leave the triangle underparameterized, as an infinity of triangles exist with the same three angles. Such parametric formulations are said to be *ill-posed*.

Incidentally, the latter is also an example of the separation of shape and scale, as defining the angles of the triangle essentially defines its shape; another piece of information is needed to give it scale. Separating shape and scale is good practice in aerodynamic shape optimization (recall Section 2.1.1), but because dealing with something completely 'sizeless' is mathematically awkward, a simple work-around is usually employed: a length related to the geometry is designated as the unit of measure for the other lengths. For example, the geometrical elements of aerofoils are usually defined in terms of their chord length. In the case of the triangle with known angles, we could set one side or perhaps the area to unity.

Item 3: Have We Created an Unnecessarily Deceptive or Complicated Objective Function Landscape?

Another key check to be applied to a particular selection of design variables must be performed in the context of the optimization problem they are to feature in. The question to be asked is: could I choose a different set of variables that would lead to a more benign objective function surface from an optimization perspective? For example, have we artificially engineered avoidable local optima?

Consider, for example, the parameterization of the lengths in a three-class cabin layout on a passenger airliner. The obvious (naive?) choice is to use the length of the individual classes: l_{1st}, $l_{business}$ and $l_{economy}$, with the overall length of the cabin being a derived quantity $L = l_{1st} + l_{business} + l_{economy}$. While this, at first glance, seems like a perfectly sensible choice, it may make the landscape to be optimized unnecessarily complicated if the objective is strongly driven by L. Since different combinations of l_{1st}, $l_{business}$ and $l_{economy}$ can yield the same L, vast numbers of spurious local optima will be introduced. This is particularly troublesome if we are aiming to perform a thorough global search of the design space. A more sensible set of parameters in this case would include L itself as a design variable, with the length fractions corresponding to two of the sections, say, $f_1 = l_{1st}/L$ and $f_b = l_{business}/L$, making up the set. This latter arrangement also makes the choice of variable ranges a little easier, partly by virtue of nondimensionalizing some of the lengths.

Item 4: 'Continuity'

This is more-or-less equivalent to the mathematical concept of function continuity (the inverted commas refer to the 'less' part), so we shall not dwell on it beyond the intuitive notion that a small change in one or more design variables ('small' to be understood here with respect to the overall range of variable) should have a small effect on the overall shape (where 'small' is to be understood in the context of overall range of shape variation that the geometry is capable of).

Failing to ensure this is likely to strain the capabilities of the optimizer unnecessarily and it makes the results difficult to interpret.

This requires particular attention when we are seeking a *robust design*. This means that we are not necessarily looking for the best global optimum, but rather for one that best balances absolute performance and robustness to small changes in shape (say, as a result of wear) or operating conditions. Such design searches become very tricky if the response of the geometry itself (to changes in design variables) can be abrupt – it will be difficult to tell whether this lack of robustness originates from the objective function or the geometry parameterization.

From an adjoint optimization point of view (more on which in Section 10.2.2), we may also consider the problem from the reverse standpoint; that is, the relationship between the location of any point of the geometry and the design variables should be differentiable.

Item 5: 'Variable Scope'

Is the scope of the chosen parameters what we intended them to be? In other words, are we satisfied that the extent of the shape affected by changes to the values of given design variables is as local/global as we had planned?

Compare, for example, Bézier splines with nonuniform rational B-splines (NURBS). Moving a control point on the latter (as in Figure 3.17) will only have a local effect – moving control points on Bézier curves will affect the entire curve. Neither type is 'right' or 'wrong' – not being clear on which type we are using *is* wrong.

2.3 What Makes a Good Parametric Geometry: Three Criteria

The checklist of Section 2.2.1 introduced some of the key features that a parametric geometry *must* have. Here, we shall review some desirable characteristics that it *should* have.

Different engineers may hold subtly different mental wish-lists under this heading, but these differences are usually in the ranking, rather than in the entries themselves. Here, then, are the top three items on *a* list – coloured, no doubt, by the authors' personal biases.

2.3.1 Conciseness

Entia non sunt multiplicanda sine necessitate – entities should not be multiplied beyond necessity. This maxim is associated with 14th-century English philosopher William of Ockham (though he appears not to have actually phrased it in this way). While at face value it appears a little banal, it is the 'soundbite' version of a rather deeper principle known as *'Ockham's razor'*. Different disciplines phrase this *law of parsimony* in different ways, but fundamentally it states that, given multiple adequate solutions to a problem, the solution postulating the fewer entities should be chosen.[4] In our case, this would translate roughly as 'of several possible

---

[4] Mathematical modelling in general and surrogate modelling in particular are enthusiastic adopters of this principle – given a set of observations, the least complex of all models capable of fitting the data should be chosen.

parametric geometries, the one with the smallest number of design variables should be chosen, all other features being equal'.

Of course, Ockham's razor is a largely aesthetic principle, but there is a much more potent, mathematical reason, too, why a geometry has to be as concise as we can make it: the size of the design search space – and, therefore, the cost of any conventional, black-box-type optimization process defined therein – increases exponentially with the number of design variables. To tackle this 'curse of dimensionality' at the outset, one must be ruthless in limiting the number of design variables. We shall revisit this notion from time to time on the pages of this book.

2.3.2   Robustness

We use the term here to describe a parametric geometry's ability, in terms of design space proportion, to yield physically and geometrically sensible shapes. A low level of robustness wastes physics-based analyses because such codes may not throw an exception immediately upon being presented with a shape that, say, intersects itself, and the result may be that such flaws only become apparent, for instance, at the end of a long mesh-generation process. By that time, minutes or even hours will have gone by, and this ultimately equates to a wasteful optimization process. Robustness can sometimes be improved by restricting design variable domains; but regions of infeasibility are seldom rectangular, so this process may sacrifice the next item on the wish-list, which is . . .

2.3.3   Flexibility

This is the breadth of the range of shapes the parametric geometry is capable of generating. It is very hard to measure, and it is generally impossible to tell when a model has reached 'sufficient' flexibility. This is because the necessary flexibility is determined by the vague and difficult to define concept of how 'unusual' would a shape have to be for it to still be worth investigating. An important point to make here is that a good strategy might be to design multilevel parameterization schemes, where additional flexibility can always be bought through the addition of further variables.

What are the practical limits of geometrical flexibility? The experience of stress-based structural shape and topology optimization would suggest that truly generic descriptions are likely to be numerical; that is, no symbolic algebra (an explicit function of the design variables) will be available. This might mean that we would have no design variables either, at least not in the conventional sense. Figure 2.3 is a 'storyboard' of the optimization process conducted on such a geometry. These shapes were generated using a technique based on the iterative migration of boundary nodes towards areas of high stress – several other nonparametric structural optimization methods are available, working on different principles (level set methods, algorithms designed to gradually 'eat away' underutilized regions in the structure, etc.).

In aerodynamic optimization, some of the so-called *free-surface-based* optimization heuristics (also known as *adjoint schemes*) come close to this type of philosophy, though they are generally unsuitable for truly global optimization,[5] as they are only really capable of

---

[5] . . . Except as part of a hybrid scheme, where a global optimizer, such as a genetic algorithm or a space-filling sample generator, is augmented by an adjoint-based multistart local optimizer.

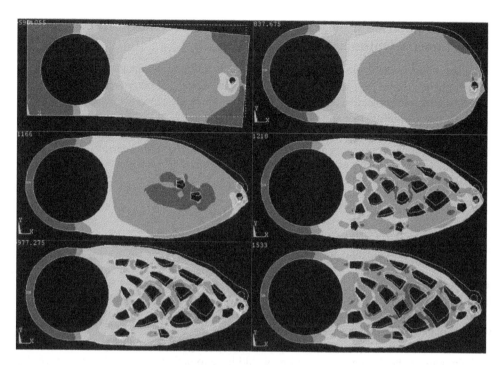

Figure 2.3 Extreme geometrical flexibility – the initial design of a cantilever bracket (top left) and five snapshots of its intermediate geometries (in a left-to-right, top-to-bottom order) in the course of an evolutionary structural optimization run. The shapes are shaded with a stress contour map.

exploiting the *basin of attraction*[6] of the local optimum nearest to their starting geometry. Additionally, the flow field variables do not provide the same sort of guidance to an aerodynamic optimization process as the stress field gives to the type of structural search shown in Figure 2.3. Luckily, surfaces wetted by flow rarely present the engineer with good reason to equip the geometry with the sort of flexibility only afforded by numerical shape definition schemes.[7]

Finally, a note on two subtly distinct forms of geometrical flexibility. Some schemes are designed to *perturb* existing, baseline shapes, thereby *adding* flexibility – this can usually be done in a cumulative way. For example, basis-function-type methods, such as the 'bump' functions of Hicks and Henne (1978) mentioned in the Preface, fall into this category – the more of these we add to the baseline shape, the greater the flexibility becomes. Other schemes have their *intrinsic* flexibility. That is, the parameterization is not built upon a baseline shape; the flexibility arises from the definition of the curve itself.

---

[6] The basin of attraction of a local minimum is the geometrical locus of all those potential gradient-descent starting points from which the optimizer will go to that local minimum.

[7] Certainly not at macro-scales – there may be an argument for trying to evolve complex shapes for, say, very small features designed to control boundary-layer behaviour.

2.4   A Parametric Fuselage: A Case Study in the Trade-Offs of Geometry Optimization

The greatest single challenge of the above wish-list is that improvements on one count can usually only be made at the expense of another. The parametric geometry construction process is often akin to multi-objective optimization – we seek the best trade-off between conciseness, robustness and flexibility. In some cases a more helpful mindset is a constrained optimization one. For instance, we may have a clear idea of how much flexibility we require and we are seeking to maximize robustness and conciseness subject to that requirement. We shall now use the parametric geometry of a generic fuselage to illustrate the tensions between the items on the wish-list.

2.4.1   Parametric Cross-Sections

A Fundamental Building Block

Let us consider the external geometry of a fuselage cross-section. In the spirit of separating shape from size, we shall merely consider the cross-sectional *shape* here, normalized to a top-to-bottom distance of one. Further, we shall assume symmetry with respect to the ZX (vertical) plane, so we only need to define half-sections.

In terms of the flexibility required of a parametric fuselage cross-section geometry, there is a certain point in the design space that one would have to include, come what may: a circle. The basic reasoning behind this stands on structural grounds. For a semi-monocoque fuselage enclosing a pressurized cabin and/or a payload bay, any shape other than a cylinder with hemispherical end-caps is a stress engineering compromise.

Although we are interested in the external surface of the fuselage, which does not always follow the shape of the pressure cabin it enclose, the external surface of the main cabin area of a large proportion of aircraft is circular (e.g. Boeing 777, nearly all business jets) – hence the need for a circular cross-section in any geometry wishing to represent the external shape of a fuselage:

$$Y(Z) = Z^{0.5}(1 - Z)^{0.5}, \quad Z \in [0, 1]. \tag{2.15}$$

This is a somewhat unusual definition for a (semi-)circle, but it offers a simple way of increasing the flexibility of the geometry – at the expense of introducing additional design variables, of course. Here is what we might term a 'generalized circle':[8]

$$Y(Z) = Z^{N_1}(1 - Z)^{N_2}, \quad Z \in [0, 1]. \tag{2.16}$$

---

[8] This is indeed *a* generalized circle – one may stumble upon other such geometrical artifices in unexpected places. One example is the modestly named *superformula* (Gielis, 2003), with its origins in botany. It has an elegantly simple formulation and its flexibility is great, though its response to parameter set changes may prove to be deceptive for optimization algorithms, and finding suitable ranges for its variables is less than straightforward. See Demasi *et al.* (2014) for an application in aerodynamic design. See also the PhD thesis of Isikveren (2002) for an alternative formulation.

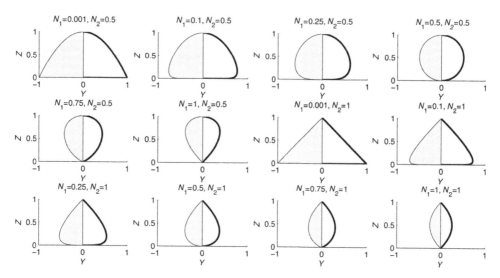

Figure 2.4 Generalized circle fuselage cross-sections obtained by assigning a range of different values to N_1 and N_2 in Equation 2.16.

By varying the two new design variables we can produce a range of possible cross-sections, as illustrated in Figure 2.4.

While this allows much freedom in shaping the cross-section, in some cases it might be desirable to add the possibility of stretching the geometry – this is a simple alteration, but one that comes at the cost of another design variable C:

$$Y(Z) = CZ^{N_1}(1 - Z)^{N_2}, \quad Z \in [0, 1]. \tag{2.17}$$

This is now a fairly flexible geometry, but it is missing two key tricks, which are particularly important in the design of passenger airliner-type fuselages. We next look at ways of addressing these shortcomings.

Increasing Flexibility

The first limitation of the model described by Equation 2.17 is that it cannot generate so-called *double bubble* cross-sections – think of the forward part of the Boeing 747 fuselage or, for less dramatic examples, the Boeing 737 family or the MD-90-30 (Figure 2.5).

Another flexibility issue is that we cannot use it to generate the wing-to-body fairing area so typical of transport aircraft.

One solution to both problems is to have two separate lobes based on Equation 2.17: the top one – let us call it the *passenger lobe* – anchored at its upper end at the $(Y, Z) = (0, 1)$ point, and the bottom one – we shall refer to it in what follows as the *cargo lobe* – anchored at its lower end to the $(Y, Z) = (0, 0)$ point. The external surface geometry will then be the

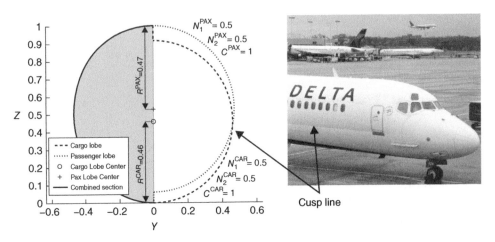

Figure 2.5 Approximation of an MD-80/90/B737-style fuselage cross-section featuring a clear cusp line between the two lobes (also highlighted on the image of a McDonnell Douglas MD-90 aircraft on the right – photograph by A. Sóbester).

maximum of these two lobes along the Y dimension; that is, the curve that envelops them from the positive side:

$$Y(Z) = \max[Y^{\text{CAR}}(Z), Y^{\text{PAX}}(Z)], \quad Z \in [0, 1], \tag{2.18}$$

where

$$Y^{\text{CAR}}(Z) = \begin{cases} C^{\text{CAR}} Z^{N_1^{\text{CAR}}} (2R^{\text{CAR}} - Z)^{N_2^{\text{CAR}}}, & Z \in [0, 2R^{\text{CAR}}] \\ 0 & \text{elsewhere} \end{cases} \tag{2.19}$$

and

$$Y^{\text{PAX}}(Z) = \begin{cases} C^{\text{PAX}}[Z - (1 - 2R^{\text{PAX}})]^{N_1^{\text{PAX}}} (1 - Z)^{N_2^{\text{PAX}}}, & Z \in [1 - 2R^{\text{PAX}}, 1] \\ 0 & \text{elsewhere.} \end{cases} \tag{2.20}$$

We now have a much more flexible geometry, one that can cover most airliner fuselage shapes at least. As an inevitable side effect we also have eight variables. This not a small number – an eight-dimensional space can be rather costly to explore. However, not every design study will require every one of these degrees of freedom. It should be easy to select the appropriate set for a particular problem, as each variable has a clear and fairly intuitive meaning.

The upper (passenger) lobe and the lower (cargo) lobe have variable 'radii' (the inverted commas refer to the ability of the functions to describe noncircular shapes): R^{PAX} and R^{CAR} respectively. The deviations from circularity are controlled, just as in the case of the basic, single-lobe fuselage section, by two exponents, or rather two on each lobe: $N_1^{\text{PAX}}, N_2^{\text{PAX}}, N_1^{\text{CAR}}$ and N_2^{CAR}. Finally, additional flexibility is enabled by a scaling coefficient on each lobe (C^{PAX} and C^{CAR}).

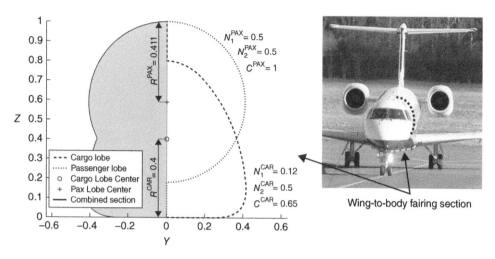

Figure 2.6 Approximation of an Embraer ERJ 145-type wing-to-body fairing area fuselage cross-section (left) with the relevant section highlighted on an image of an ERJ 145 aircraft on the right (photograph by A. Sóbester).

Figure 2.5 depicts an instance of the parametric model described by Equation 2.18: a slightly eccentric double-lobe fuselage cross-section, similar to that seen on the Boeing 737 series of airplanes or on the MD-80/90 family. The values of the eight design variables are also indicated on the plot.

Figure 2.6 is a further illustration of the flexibility of this formulation, showing how it can model a wing-to-body fairing section similar to that featured by the fuselage of the Embraer ERJ 145 regional jet.

The same equation is used in Figure 2.7 to construct a Boeing 747-style double-deck fuselage cross-section. To enable an optimizer to include this type of section in a design search, we need to make a subtle tweak to the formulation: we need to add the possibility of closing the profile with a common tangent (shown here with a dash–dot line). If we do not want this to be a permanent feature of the geometry (i.e. we also want the optimizer to visit solutions that leave the cusp between the lobes exposed), we have to pay with the usual currency of flexibility enhancements: increased dimensionality. Thankfully, this time we can get away with a binary variable (common tangent to the two lobes present or not); but even this results in a doubling of the design space, and thus a doubling of the cost of the design search – not a step to be taken lightly. Additionally, this change might raise a warning under item 3 of the checklist of Section 2.2.1 – a deep cusp becoming exposed or concealed by the straight-line segment might yield a discontinuity in the objective function.

An alternative here might be to embed a rule into the geometry formulation that automatically adds the tangent if the cusp is too deep, thus dodging the additional design variable. Of course, a side effect of this move is a subtle loss of flexibility (we have removed deep cusp shapes from the design space). In any case, the more general point here is that the more design rationale we can equip the geometry formulation with, the smaller and more manageable the design space is likely to become. Scripting-based geometry models hold an implementation advantage here

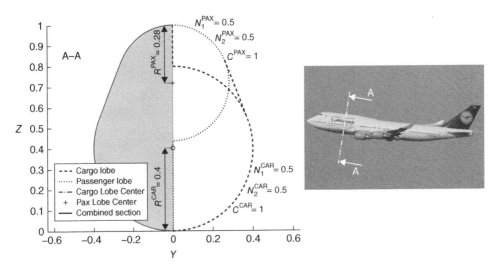

Figure 2.7 Boeing 747-style cross-section geometry as an instance of the model of Equation 2.18 (photograph by A. Sóbester).

as they naturally admit the embedding of such rules, though most interactive geometry engines also have some features geared towards this type of design thinking.

2.4.2 Fuselage Cross-Section Optimization: An Illustrative Example

Let us now test the flexibility of the formulations discussed above on a simple optimization problem. To enable the reader to experiment with variations on this design case study (for which the code is provided in the toolkit accompanying this book) we shall use an objective function that is based simply on geometrical calculations, and it is thus very quick to compute.

Consider the problem of designing a cross-section geometry for a single-aisle aircraft with a six-abreast seating configuration and sufficient space in the cargo hold for a standard LD3-45W unit load device – a cargo container commonly used on the Airbus A320 family of aircraft. This simple specification translates into a large number of geometrical constraints – Figure 2.8 shows the representative set that we will take into account in this study.

The '+' symbols on the sketch indicate the boundaries of regions that the cabin wall must not protrude into. We have defined these on the basis of seating a 95th percentile US male passenger at the window seat and ensuring adequate clearances around his head when standing in the aisle, when stepping under the overhead bins and when seated, as well as around his outside shoulder, elbow and foot when seated.

Figure 2.9 is a histogram of seat widths across the world's larger airlines, showing that the bulk of the field is clustered in the 17–18 inches band – for the purposes of this case study we split this band in half, using a width of 17.5 inches (0.44 m). The reader may wish to experiment with changing the seat widths – perhaps to gauge the fuel burn penalties of added comfort, one of the important battlegrounds in the passenger aircraft industry.

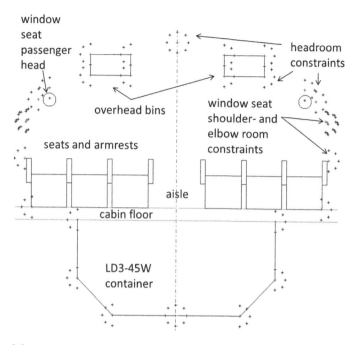

Figure 2.8   Internal geometrical constraints on the shape of a passenger airliner cabin.

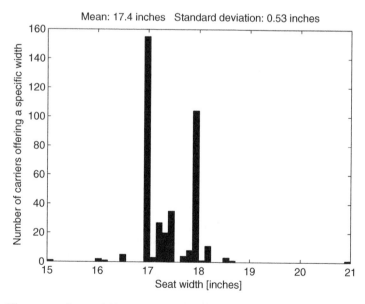

Figure 2.9   Histogram of seat widths across a sample of the world's larger airlines (data courtesy of seatguru.com).

Additionally, we have reserved a fixed depth for the floor beam, as well as space around the edges of the container and the boundaries of the protected areas mentioned above to accommodate the (assumed fixed) depth of a fuselage frame. '+' symbols mark all of these hard constraint zones. The goal of the exercise is therefore to wrap these markers with the 'tightest' possible cross-section curve.

In what follows we shall take a three-stage approach to solving this problem, gradually ramping up the flexibility of the cross-section geometry.

Stage 1: A Circle

From a purely structural standpoint, the most efficient shape for a slender pressure vessel is a cylinder with hemispherical end caps. As a result, a large proportion of aircraft with pressurized cabins feature a fuselage with a circular cross-section.

We therefore begin our investigation by assuming that the shape of the section is fixed and it is defined by Equation 2.15. With the shape thus settled, the only other element we need for this optimization exercise is a uniform scaling variable. Technically, we also need to vary the vertical position of the scaled (semi-)circle placed around the constraint points, but this problem can, to some extent, be separated from the main optimization problem, so we could refer to the latter as a one-variable problem, with the proviso that each tentative solution has to undergo a simple additional search (a kind of repair step) to find its optimum position (this will be the case for the next two stages of the search too).

Therefore, the problem is to minimize the cross-sectional area objective function associated with the circle, as a function of a single (scaling) variable, subject to the hard constraint that all the '+' markers have to be enclosed by it. From an optimization perspective the problem is a trivial one (requiring a simple line search) and, as shown by the objective function history of the search depicted by the thick black line in Figure 2.10, it is pretty well converged in about 60 evaluations of the objective (the history curve turns very shallow at that point, with some very fine tuning taking up another 300 evaluations or so). The sketch labelled 'circle' shows the resulting geometry.

Stage 2: A 'Generalized Circle'

Time to make the geometry more flexible. We might ask: what gains could we make on the cross-sectional area (a surrogate, in essence, for the pressure drag of the fuselage) at the cost of sacrificing some of the structural advantages of a circle? This means throwing more variables at the problem, as we move on to Equation 2.17: we make the two exponents N_1 and N_2 free to vary, as well as adding the 'sideways stretch' variable C. As before, we also have the uniform scaling variable and the quasi-separable vertical adjustment variable.

This is a slightly more serious optimization challenge, and we used the pattern search of Nelder and Mead (1965) to accomplish it. Starting from the circle ($N_1 = 0.5, N_2 = 0.5, C = 1$) we arrive at the geometry labelled 'generalized circle' in Figure 2.10.

The greater number of variables (three for the shape definition, one for scale and one separate adjustment variable) means that convergence now took longer (around 200 objective function calculations), but the impact of the increased flexibility is also clear: we have improved upon the cross-section area by a significant margin (the search history is shown by the dotted line in

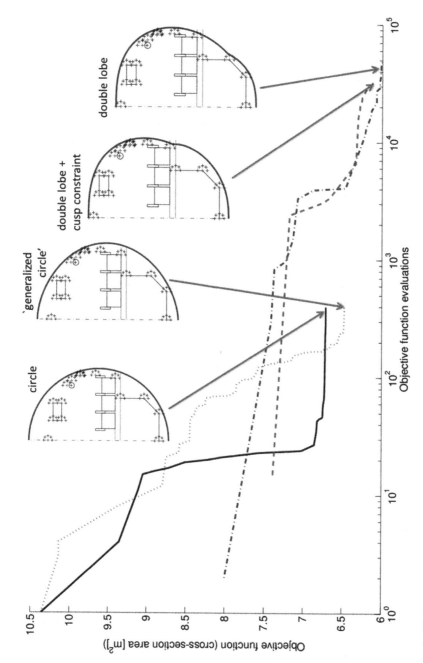

Figure 2.10 Higher parametric geometry dimensionality: greater flexibility and better results (smaller ultimate cross-section area), but the optimization can be orders of magnitude more costly.

Figure 2.10). Note, however, that the 'stiffer', circle-based search was 'ahead' for the first 10 or so evaluations, indicating that, when on a very tight budget, it is best to keep things simple.

Stage 3: A Double-lobe Geometry

Time to move on to Equation 2.18, our most flexible cross-section geometry. The number of shape variables doubles here, as we have now introduced a new set for a separate cargo lobe. We also have the R variables here, which determine the relative positions of the two lobes.

The impact of this increase in flexibility is, as seen in Figure 2.10, twofold. First, the cost of the optimization process (now conducted via a genetic algorithm, followed up by a Nelder and Mead pattern search to fine tune the best solution found by the genetic algorithm) has gone up by two orders of magnitude – we are into the tens of thousands of evaluations of the objective function. However, there is a positive, second impact too: we have succeeded in the main point of the flexibility increase; that is, we have managed to drive the cross-section area much further down. The corresponding geometry is labelled 'double lobe' on the right-hand edge of Figure 2.10.

Stage Three with an Additional Constraint

We now have a tightly wrapped set of constraint points, though we got there at a high cost – the optimization problem has become very expensive. Given the simple, analytical test function, this was a barely noticeable minor inconvenience for this simple problem, the search taking a few minutes, as opposed to a second or two in stages 1 and 2. However, the increase would have been immensely inconvenient if a high-fidelity numerical analysis process (perhaps involving multiple disciplines) had been needed to compute the objective.

Also, this geometry now has a blemish: the cusp between the two lobes is nowhere near the cabin floor level – this is likely to be a structurally inefficient solution. We thus reran the search process with the same geometry model, the same objective and the same optimizers, but we have added another hard constraint: the cusp has to line up with the floor beam.

The result, as illustrated by the last but one image in Figure 2.10, is a neater-looking design, but a slightly greater cross-section area.

Postscript

The key message of the example above is that the dimensionality of a parametric geometry (and thus its flexibility) has a powerful impact on the associated optimization process; and because the effective optimization of some measure of merit linked to the geometry is practically the only reason why one would build a parametric geometry,[9] this is a very important conclusion.

Of course, the 'curse of dimensionality' can, in some cases and to some extent, be mitigated by clever optimization strategies. For instance, the choice of the starting geometry (or the population of starting geometries, in case of an evolutionary algorithm) might reduce search time (though, if done without due care, it might also narrow the design space, precluding unexpected, novel, disruptive solutions).

---

[9] Note that we are using 'parametric' in the 'shape depending on design variables' sense.

Also, sometimes low computational cost objective function gradients are available and the calculation of these may scale favourably with the number of design variables (we have already hinted at the favourable characteristics of adjoint solvers, and we will take a close look at their use in such design processes in Chapter 10). To some extent, this relieves the pressure on dimensionality, but the overall cost of the optimization problem will still increase with increasing dimensionality. Whatever the choice of design exploration technique, *entia non sunt multiplicanda sine necessitate*...

2.4.3   A Parametric Three-Dimensional Fuselage

Having used the increasingly flexible fuselage sections to give the reader a sense of the cost of flexibility, we shall now complete this fuselage design example by extending it into the third dimension, adding yet more degrees of freedom. We shall do this through a sort of 'extrusion' along the X-axis. Looking at it another way, we need to construct a series of two-dimensional (2D) sections spread out along the X dimension and loft a surface over it.

Whichever way one views this step, though, in essence we now need to regard the variables of the cross-section geometry as parametric functions themselves. Consider Figure 2.11 for an example: the geometry of a fuselage similar to that of the Embraer ERJ 145 regional jet.

In Figure 2.6 we see the instance of our double-lobe cross-section geometry that might produce the central wing-to-body fairing area. The question is, what longitudinal variation of the parameters would produce the transition from the circular section to this double-lobe section? These are the variations shown alongside the fuselage in Figure 2.11.

With a relatively simple forward and aft section geometry, the cargo lobe of our parametric geometry is only needed for the wing-to-body fairing area, so N_1^{CAR}, N_2^{CAR} and R^{CAR} are only defined in this portion and the cargo lobe 'sideways stretch' coefficient is set to zero elsewhere (with a constant value of 0.65 in this region). The streamwise variation of the passenger lobe parameters features short, smooth deviations from the baseline values corresponding to the circular cross-sections of most of the fuselage. One such deviation is N_1^{PAX} falling away gently from 0.5 to 0.43 near the front to yield the slightly 'squashed' shape of the nose section.

These variations now define the cross-section *shapes* – all that is left to do is to establish the uniform *scaling* parameter, as well as the appropriate vertical positioning parameter. Again, these now become functions (of a streamwise station variable), and together they basically define the side profile of the fuselage. Put another way, they can be derived from the two curves (upper and lower) making up the side profile of the fuselage: the scaling variable is the difference between the two and the Z-wise positioning variable is the lower curve. The particular instance of the pair of curves that was used to generate the fuselage model of Figure 2.11 is shown in Figure 2.12.

Any number of mathematical formulations could be used to define all of these curves, subject to any number of design variables. In Figure 2.11 we merely show a possible instance of one possible parameterization. There is no limit here to adding additional flexibility via the parameterization of these streamwise functions. A desire to produce a stealthy shape, to blend in a conformal fuel tank or perhaps the need to apply transonic area ruling to a portion of the fuselage, may demand significant amounts of flexibility. However, once again, the curse of dimensionality may limit how much flexibility we can actually afford. We will have more to say on this topic in Section 2.6.

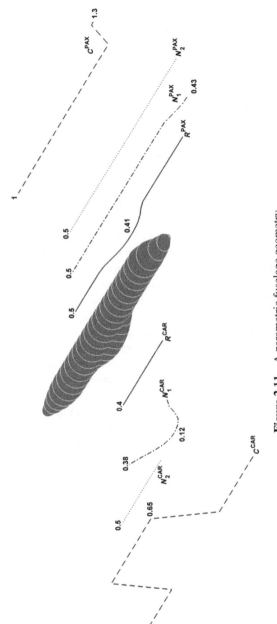

Figure 2.11 A parametric fuselage geometry.

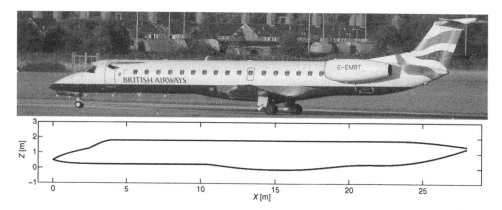

Figure 2.12   Side profile curves used in the definition of the fuselage geometry shown in Figure 2.11 (loosely based on the fuselage geometry of the Embraer ERJ 145 shown above – photograph by A. Sóbester).

2.5   A General Observation on the Nature of Fixed-Wing Aircraft Geometry Modelling

The external geometry of a topologically simple three-dimensional (3D) body can be viewed as a surface lofted over a series of cross-sections generated by parallel cut planes. The fuselage parameterization scheme discussed above is an example of this philosophy, and to illustrate this, in Figure 2.11 we highlighted such a set of cross-sections (white curves set against the grey surface) on an instance of this parametric geometry.

This is not the only way of regarding the geometry of a topologically simple body – it is, for instance, not the way in which a sculptor would see the shape of a clay human head taking shape under their hands. It is, however, a very convenient way of capturing a shape from an engineer's point of view, and this is more than a throwback to the pre-computer-aided design (CAD) days when 2D drawing boards offered no real alternative. It is an especially convenient way of describing a geometry when there exists a cutting plane orientation that either makes design constraints very clear (as in the case of the cabin cross-sections discussed above) or the physics more intuitive – the stresses in pressure cabin skins or aerofoil sections through wings are good examples.

What about topologically more complex bodies? The engineer's instinct is to divide and conquer; that is, to break a shape down to primitives that can be conceptualized as a loft over sensible section curves. Conventional fixed-wing aeroplanes are very amenable to this way of thinking, at least at the conceptual or early preliminary design level. The classic passenger airliner geometry, for example, is a gift to those thinking in this way (perhaps there is an interesting cause and effect argument to be had here). It is usually made up of a larger and several smaller bodies lined up with the flow (the fuselage and the engine nacelles) and two larger and several smaller bodies positioned roughly perpendicularly to the flow (wings, tailplanes, fin).

Not every branch of engineering design (or, indeed, aircraft design) is this well suited to this philosophy. Revisit the structural geometries of Figure 2.3 for an example where this way of thinking does not help at all, chiefly because shape and topology are no longer clearly separable. There may come a point when aerodynamicists will feel the need for a comparable amount of shape- and topological-freedom, but, for now, a surface lofted over 2D sections is the *status quo* of the industry, and it is the philosophy adopted in this book too.

We exemplified the above idea in this chapter through the problem of fuselage modelling. We showed that we can start with 2D sketches and move into the third dimension by introducing a series of functions of a streamwise variable (x is an obvious choice) that describe the shape and positioning of the parametric cross-sections, over which we can then loft a surface (in fact, if these streamwise functions yield a value for any X, they define the 'loft' too), as well as a profile section that determines the scaling and positioning of the support curves of the loft.

In Chapter 9 we shall discuss the construction of 3D wings in the same spirit, building upon the parameterization of aerofoil cross-sections (Chapters 5, 6 and 7) – analogous to the fuselage cross-sections here – and planforms (Chapter 8), analogous to the side profile definition of the fuselage in this chapter.

2.6   Necessary Flexibility

One of the key themes running through this book is the control of the dimensionality of a parametric geometry – that is, how we make sure that the desire for flexibility does not render the geometrical formulation completely unusable in an optimization context.

Let us return, for the moment, to the parametric fuselage cross-section model of Section 2.4.1. We gradually built up a geometry capable of producing a wide range of shapes, including double-lobe geometries. The optimization exercise we performed on this geometry highlighted the rate at which the cost of searching the corresponding design space shot up as we added more variables (Figure 2.10). And yet, theoretically, the scope of this family of shapes can be expanded further still, rather readily.

Equation 2.16 is, in fact, one of the *class functions* introduced by Kulfan (2008) as part of the *class-shape transformation* parameterization scheme. This formulation, which we will return to in detail in Section 7.3 (precisely on account of its impressive flexibility), allows the definition of a *shape function S*, which broadens the scope of the basic single-lobe geometry of Equation 2.16 to a virtually infinite range of shapes of the form

$$Y(Z) = S(Z)Z^{N_1}(1-Z)^{N_2}, \quad Z \in [0, 1]. \tag{2.21}$$

As we will see later, these shape functions are typically sums of polynomial basis functions, and the greater the number of these bases, the more flexible the geometry becomes – naturally, at the cost of additional design variables.

Equation 2.21 could thus be turned into what is a virtually universal approximator of fuselage sections (or, indeed, sections of other bodies, such as engine nacelles, pods, fuel tanks, etc.). But, each time we add another shape function into the mix, we must ask if the additional flexibility is really necessary. Referring back to Section 2.3.1, we must remind ourselves of the curse of dimensionality and not forget that the cost of design optimization increases exponentially with the number of design variables.

Considering, for instance, a sampling of a design space through a full factorial plan (design of experiments) with three observations along each dimension, this will cost us nine experiments in two dimensions, 27 in three and so on. Each new variable triples the cost of exploring that design space. So, by the time we get to the tenth variable (basically the dimensionality of our cross-section geometry, once we have included scaling and positioning, but before we even thought of shape functions, as in Equation 2.21!) we need $3^{10} = 59\,049$ analyses to maintain the same observation density, and adding just one more (one polynomial basis function, perhaps?) takes us to $3^{11} = 177\,147$. And this is just for three levels per dimension, which does not even seem that much. After all, to step back to a more intuitive dimensionality, a complicated response is not that easy to reconstruct from nine experiments in two dimensions!

Having to fit into sensible computational budgets usually demands that we abandon full factorial designs, but whatever experimental set-up we might choose (e.g. a Morris–Mitchell optimal Latin hypercube or some other space-filling plan) the cost still rises exponentially with the number of design variables.

Once again, *entia non sunt multiplicanda sine necessitate...*

2.7 The Place of a Parametric Geometry in the Design Process

2.7.1 Optimization: A Hierarchy of Objective Functions

In Section 2.1.2 we introduced some typical, simple objectives one may wish to optimize as part of the aerodynamic design process. These slot in at the lower levels of a hierarchy of objectives usually associated with most aerospace programmes.

At the top of this hierarchy one may find objectives such as life cycle cost or profit; but since these are usually very hard to connect to the design variables via objective functions, lower level related objectives may be used. Fuel burn and perhaps range are typical of this second tier.

These are, however, still somewhat difficult to link directly to outer mould line geometry variables, so another level is required. Sometimes these lower level variables emerge from simple analytical models of the second-tier variables. A classic example is range (R), an initial estimate of which can be obtained from the *Breguet range equation*:

$$R = \frac{\lambda}{c}\frac{C_L}{C_D}\ln\frac{W_i}{W_f},\tag{2.22}$$

where λ is propulsive efficiency, c is specific fuel consumption, C_L is the lift coefficient and W_i and W_f are aircraft weights at the beginning and the end of the cruise segment respectively.

We can immediately see a term there that makes an ideal optimization objective for the aerodynamic designer: C_L/C_D. If a flow simulation is in place, which is capable of computing this ratio (in the form of a complete drag polar or perhaps drag for a given lift), we have now identified a manageable aspect of the very difficult high-level problem and the optimization of C_L/C_D will contribute to optimizing range for a given amount of fuel, as well as the top-level objective.

Of course, the other terms of the range equation will have their own optimization problems associated with them, and there may be many other such objectives on the other branches of the hierarchy. There may also be constraints, which limit the design variable domains within which the optimization algorithms can operate.

2.7.2   Competing Objectives

A further complication is that sometimes a group of such objectives, linked to the same geometry parameters, are in conflict with each other. The solution in such cases is not a single design variable vector that maximizes or minimizes an objective, but rather a hypersurface of designs, which represent equally valid compromises between optimizing the various objectives. These are referred to as *Pareto surfaces* or *Pareto fronts*.

A Simple Example from Aerofoil Design

Here is a very simple Pareto problem centred around a parametric geometry. We need to design an efficient wing, which will also serve as a voluminous fuel tank. The latter requirement would clearly tend to drive up the thickness of the wing section, while the former would tend to drive it down – a classic case of tensions between two objectives.

From the flow conditions and the overall estimated weight of the aeroplane we can compute the required aerofoil lift coefficient c_l, so we can perform an iterative flow analysis around each candidate design until we find the angle of attack that yields that target c_l. A simple (low-level) efficiency objective can then be the drag coefficient obtained at that angle of attack.

A suitable aerofoil shape parameterization needs to be identified next (we will consider plenty of options for this in Chapters 6 and 7), as well as a 2D viscous flow simulation code capable of the calculation outlined above (a simple 2D panel code with viscous boundary layer, implemented in MATLAB® and Python is included in the toolkit accompanying this book, and is employed in the human-powered aircraft case study in Chapter 12).

There is a rich literature on the efficient identification of the Pareto front of two objectives of a design problem like this, but here let us assume that the thoroughness of the search trumps efficiency because the solution process is very quick. The 2D drag calculations using full potential codes with viscous corrections only take a few seconds per design, and we will simply use maximum thickness as a surrogate for fuel tank volume – this takes an insignificant amount of time to compute. A 'brute force' technique is therefore to spray the design space defined by the variables of the parametric aerofoil with a uniform, dense coverage of designs – let us say, using a Latin hypercube sample planning algorithm (McKay *et al.*, 1979). The two objectives can be calculated for all of these designs, and the boundary of the cloud of points thus obtained will be the Pareto front – Figure 2.13 illustrates this.

The Pareto front of this cloud of points (coloured according to the camber of the aerofoils) is the top-left boundary, as we are aiming to maximize thickness and minimize c_d. The Pareto optimal, or non-dominated subset of this large set of designs is highlighted by black circles. This set is selected such that *any other selection that would lead to an improvement against one objective, would lead to a deterioration against another.*

The designs highlighted with black circles (we have also plotted the corresponding aerofoil alongside some of them) are therefore all optimal in the Pareto sense and we need another objective or some design constraint to pick the final design. We thus may end up choosing any of these geometries off the front; but, assuming that we trust the analysis and on the available information alone, there is no good reason for choosing any other designs from the cloud – see Sóbester and Keane (2007) for a more detailed look at this problem.

Incidentally, the design study depicted in Figure 2.13 was based on an aerofoil parameterization using *Ferguson splines* – we shall discuss this in detail in Section 7.2. For now, let us consider another case of Pareto analysis, this time built around a 3D geometry.

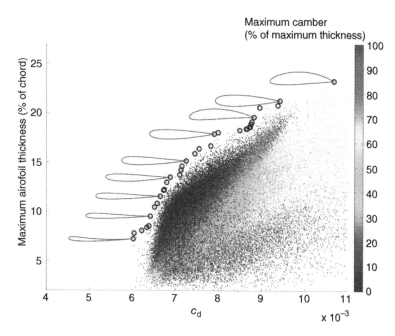

Figure 2.13 The two objectives, c_d and maximum thickness, corresponding to just over 130 000 aerofoils generated using a parametric model and a space-filling experiment planning algorithm. The non-dominated points are highlighted with black circles and the aerofoils they represent are also shown alongside some of them.

A More Complex Example: Propulsion System Integration

Consider a more complex multi-objective study, this time with more than one discipline providing the objective function values. Consider a passenger airliner with engines installed above its wings, as illustrated by two example geometries shown in the top left-hand corner of Figure 2.14. There are numerous advantages to such an installation (see Powell *et al.* (2012) for a comprehensive list), key amongst which is that the wing shields the broadband fan noise of the engine from communities on the ground.

The question arises as to where exactly above the wing would one position the nacelle. Conventional under-wing installations do not offer much freedom in this respect (the solution is usually up against the ground clearance constraint), but here we can play with two variables: the streamwise station of the intake lip x and the vertical position of the nacelle centre line z – both shown on the sketch in the top-right corner of Figure 2.14.

There are a number of possible objective functions we could consider here. For the purposes of this example, let us look at three: overall airframe drag C_D, noise shielding Δ and DC(60), a measure of the pressure distortion coefficient at the fan face, expressing the contrast between the mean total pressure over a 60° sector and the mean total pressure across the entire fan face. The two plots in the lower half of Figure 2.14 illustrate the trade-offs between these three objectives. Clearly, some regions of these objective spaces can be discarded, as designs can be found elsewhere on the surface that provide better solutions against all objectives. There are, however, conflict regions too – for instance, looking at the bottom right-hand fringes of

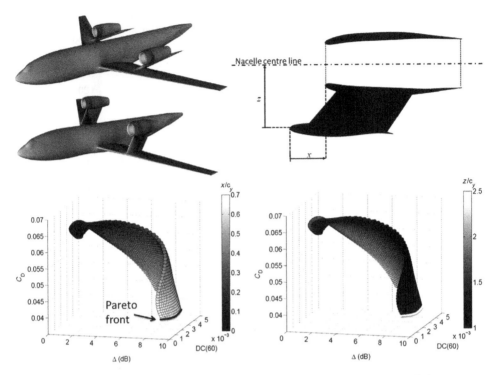

Figure 2.14 Triple-objective optimization problem based on an over-wing engine installation geometry with two design variables. Counterclockwise from top right: parameterization sketch, two extreme example designs, Pareto surface coloured by normalized x value and Pareto surface coloured by normalized z value – figure based on Powell (2012).

the two plots, we can see a 'fold' illustrating the fact that in the low airframe drag region there is competition between Δ and DC(60) – this is where the Pareto analysis becomes necessary (we have highlighted this Pareto front on the figure). As per the earlier definition of Pareto optimality, one of these designs could only be replaced by another at the price of deterioration against at least one of the objective axes.

A Key Limitation

A key limitation of the Pareto approach is the curse of dimensionality in a slightly different guise. When the number of competing objectives exceeds three or four, building the Pareto trade-off surface becomes rather expensive. Moreover, it becomes next to impossible to visualize in an intuitive manner.

A typical case of the curse of dimensionality precluding Pareto analysis in aircraft design is the treatment of multipoint cases. These commonly appear when multiple flight conditions have to be considered. A classic example is having to optimize cruise and holding/loitering fuel burn simultaneously, each at several points during a mission with different fuel and payload weights, as well as at a range of different density altitudes. The multiplicity of off-design conditions is a particularly pressing concern in the case of high-speed transports, where sets of design conditions can feature significantly different Mach numbers (Cliff *et al.*, 2001).

While the mathematical formalism of the problem is simple – a compound objective function is usually generated as a linear combination of sub-objectives corresponding to the various design points – the solution is not. The difficult part is selecting the appropriate weight for each term of the linear combination. There is no silver bullet solution at present, though there are some promising lines of research (e.g. a method involving weights that adapt to the problem as the optimization progresses, developed by Buckley *et al.* (2010)).

2.7.3   Optimization Method Selection

A modern design process relies on a parametric geometry and a means of simulating some aspect of the operation of the aircraft it models in order to obtain a measure of merit (objective function). While this is a useful combination in itself (e.g. it enables variable sensitivity studies), a truly effective design process requires that we couple it to an optimization algorithm. Depending on the computational cost of the simulation, as well as the number of design variables and the likely shape of the objective function landscape and the scope of the search, a number of classes of methods are available. Here is a brief overview of these, noting that this is one of several possible classifications and there are overlaps between the classes – they are thus best treated simply as labels one might attach to a particular algorithm.

Local Methods

Local optimizers are also sometimes referred to as *hill-climbers*, a term that aptly describes their mode of operation: starting from an initial design, they iterate towards the local maximum (although, in fact, engineering problems usually involve finding a minimum, so perhaps, 'hill-descenders' would be a more appropriate term), the basin of attraction of which the initial design is located in. Some hill-climbers require objective function gradient information, others do not. *Conjugate gradient* techniques fall into the former category, and they are available in a range of iteration step length and direction calculation methods. One of the most efficient and robust such heuristics is the *Broyden–Fletcher–Goldfarb–Shanno* (BFGS) algorithm (Broyden, 1970). These techniques really come into their own when the objective function sensitivities with respect to the geometry parameters can be computed efficiently and accurately – we dedicate Chapter 10 to objective function derivatives and their computation in an aerodynamic optimization context.

When sensitivities are not readily available, local optimization is still possible – pattern search algorithms, such as those by Hooke and Jeeves (1961) and Nelder and Mead (1965) are amongst the best known. These are also generally more robust to 'noise' that may be corrupting the objective function landscape.[10]

---

[10] The inverted commas around 'noise' allude to the fact that, unlike the noise of physical experiments, this type of computational 'noise' is deterministic; that is, running the simulation code twice will result in the same systematic error.

Global Methods

Most aircraft design engineers will agree that few nontrivial design changes are truly local. Consider, for instance, the case of the Boeing 747-8 wing design. The original main wing aerofoil was replaced with a deeper supercritical section. This caused the wing weight to increase. The centre of gravity shifted accordingly, which meant that parts of the tail had to be redesigned to satisfy stability constraints. In turn, this led to further changes elsewhere. As a result, the aircraft fitted with the new wing has 15% lower fuel burn than the 747-400, but the design effort extended far beyond the wing cross-section actually responsible for this improvement (Trimble, 2013).

In the earlier stages of the design process, however, even manifestly global changes are less disruptive, and that is the time for a *global design search*, when the number of active design variables is large and their ranges are broad.

When wholesale design edits are acceptable, the objective function landscape can be expected to be *multimodal* (i.e. to have multiple local optima) – this makes for a harder, more expensive optimization process. Local search methods may still be deployed at this stage, but they require multiple restarts from a range of starting geometries. These starting geometries should fill the design space relatively uniformly, increasing the probability that we exploit the neighbourhoods of all radically different designs.

If the cost of computing the objective function is relatively low and the number of design variables is high, evolutionary optimization methods can make very efficient global optimizers. A large range of such methods is available, all sharing the fundamental philosophy of imitating the mechanics of natural evolution. Natural life results from the nonrandom survival of randomly varying replicators – in evolutionary optimizers (such as *genetic algorithms*, which are the best known representative of this class), encoded versions of design variable strings take the role of the replicators, and the objective function, through some artificial selection scheme, determines survival (or loss) of candidate geometries.

Evolutionary algorithms typically operate over a discretized version of the design space, so solutions resulting from them may need to be fine tuned via more fine-grained local search methods (occasionally these are integrated more tightly into the evolutionary algorithm through a process imitating the idea of *Lamarckian learning*).

Surrogate Model-Based Methods

The simulations required for the calculation of objective functions may take up computational resources that limit such calculations to numbers that are too low for any conventional global search technique to make much progress with. The alternative is to fit a *surrogate model* of the objective function. This is a cheap-to-compute statistical model that aims to emulate the real objective function by learning its behaviour from the few simulations available. Clearly, the surrogate will not be able to capture *all* the behaviour of the physics-based simulator, but it may capture *enough of it* to guide an optimizer to at least the neighbourhood of the global optimum.

Specialized surrogate-guided optimizers are also available, which, through a Bayesian update scheme, iteratively enhance the accuracy of a surrogate in promising regions; that is, regions of the objective function landscape that are either poorly resolved (i.e. we are highly uncertain of their shape) or are likely to harbour a better optimum value than those found up to that point (Jones *et al.*, 1998). In Forrester *et al.* (2008) we provided a practical introduction to the use of these methods in engineering design.

2.7.4 Inverse Design

The optimization philosophy discussed up to this point is based on the idea that we seek the set of parameter values defining the geometry that minimizes or maximizes some overall objective(s). A typical way of obtaining such an objective value is to integrate it out of the flow field obtained from a computational simulation – drag for a given amount of lift is one such number.

An alternative way of regarding geometry optimization is to *seek the set of parameter values yielding the geometry that will produce a specified set of flow field features*. In Section 10.4 we will take the reader through a case study in *inverse design* of this type. For now, let us take a higher level look at its place in the aircraft engineering process.

A typical early example of this approach was the design of wing sections for transonic and supersonic flight, where the aim in the 1960s and 1970s was to create an aerofoil with a relatively flat upper pressure profile, which was to provide for a weaker shock relatively far aft (see Figure 6.8 for such a pressure profile and Section 6.3 for a more detailed discussion of its design rationale).

Of course, the ultimate aim was still to reduce drag – specifically, to engineer a delayed and slightly less aggressive transonic drag rise – but the inverse design approach provided the aerodynamicist with a more intuitive way of gauging the effect of geometry modifications

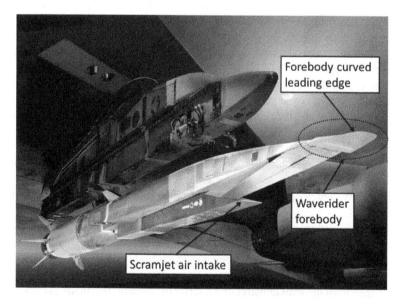

Figure 2.15 Boeing X-51A waverider attached to a pylon under the wing of its Boeing B-52 mothership. A key geometrical feature of such vehicles is the shape of the leading edge of the forebody, which will hold the shock wave in hypersonic flight and will define its shape (image courtesy of the US Air Force).

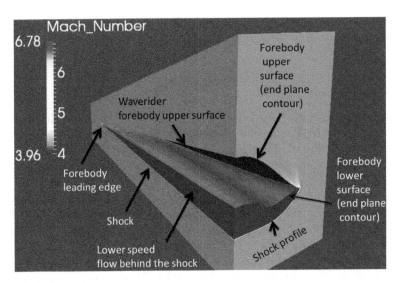

Figure 2.16 Starboard-side half-section through the forebody of a Mach 6 waverider and a block of the flow field surrounding it. A commonly used design algorithm involves a combination of standard optimization and inverse design.

(especially when the 'geometry' was represented by a wind-tunnel model) and the impact of target pressure profile changes on the overall objective was relatively well understood.

Modern inverse design truly comes into its own when the design space is very large – too large for 'straight' objective function optimization around a parametric geometry. In such cases a hybrid approach may be the most efficient way of getting a reasonable answer. Some target flow field feature may be used either for the inverse design of part of the geometry or as a starting point for a subsequent objective-function-driven search process.

Consider the example of designing a *hypersonic waverider*, such as the X-51A shown in Figure 2.15. The shock wave generated by a waverider is attached to the carefully shaped leading edge of the forebody, thus capturing the aft-shock high-pressure area underneath the vehicle. This ensures efficient lift generation, as the aircraft effectively 'rides' on its own shockwave.

A hybrid inverse design technique for generating forebody geometries was introduced by Sobieczky *et al.* (1990). The so-called *osculating cones method* involves creating a 2D parametric geometry, which defines the end plane contour of the upper surface of the forebody (see Figure 2.16).

At the same time, and this is the inverse design component of the algorithm, the desired shock profile is defined, also in the end plane. This is the propulsion system engineer's input into the process, as the operation of the scramjet intake is highly dependent upon the geometry of the shock.

The design algorithm involves constructing a series of 'kissing' (osculating) circles attached to the shock profile (in the end plane), each with a radius equal to the curvature of the shock profile in the contact point. These circles form the bases of a series of cones (whose angle is a design variable) around which the inviscid flow (described by the Taylor and Maccoll (1933) equations) is computed. Next, the upper surface profile is projected forward until it meets the shock surface and, in the osculating plane, the points of the leading edge are traced back to the end plane along the streamlines of the cone flow. The intersections between these streamlines and the end plane will trace the lower surface of the forebody.

The advantage of the method described above is that it provides a fast way of generating a forebody geometry that meets our flow field requirements (in this case in terms of the shape of the shock surface) – this is the inverse design part – while also allowing the process to be integrated into an objective-function-driven optimization loop that manipulates the design variables of the trailing edge. The objective function of the latter could be computed, for instance, through a more detailed, expensive, high-fidelity simulation of the viscous flow around the forebody or even the complete vehicle.

This hybrid technique compares favourably with a purely objective-function-driven approach, with a flexible parameterization of the entire forebody shape (not just the upper surface contour), which would, no doubt, yield a design space comprising very large sub-regions where the shock would have the wrong shape or would not stabilize on the leading edge. This would be difficult to detect prior to the analysis, and therefore large computational resources would be wasted on evaluating infeasible geometries.

3

Curves

Chapter 2 considered the kinds of traits that should be possessed by a geometry intended to form part of an optimization process, but, with the exception of illustrative examples, we have yet to look at how to actually construct such geometries; that is, the step between some design parameters and a shape whose performance can be analysed (and so optimized). Now we will look at some fundamental building blocks en route to creating parametric geometry. Naturally, we begin with curves in this chapter before introducing surfaces in Chapter 4. While the fundamental curves detailed can be produced in commercial CAD tools, we have included MATLAB® and Python code in the text and the accompanying toolset with a view to providing both greater understanding and also the possibility of the reader incorporating elements in their own bespoke toolsets. Here, and increasingly throughout the book, we also include implementations in OpenNURBS/Rhino-Python.

3.1 Conics and Bézier Curves

The classical,[1] and initially intuitive, definition of conics is that of a plane intersecting a cone, with the angle of intersection determining the shape. Here, we hope to give more insight and intuition, particularly when considering parametric conics and then the more complex NURBS, by using the projective geometry definition. We will only cover the essentials of projective geometry, and the reader wanting more on in this area may wish to refer to Farin (1999), which inspired this section.

Figure 3.1 shows a conic as the intersection of lines. The conic is defined by two tangent lines (shown in bold), here through $a^{(0)}, a^{(1)}$ and $a^{(1)}, a^{(2)}$ (but could easily be any two of the lines shown), and a third *shoulder tangent* (also shown in bold). Described in this way, the conic is a mapping or *projectivity* between the two tangent lines, with this projectivity defining the shoulder tangent. The next section covers this projective geometry construction of conics. Those wishing to get some geometry on the screen straight away might wish to skip some of the theory, go straight to Equation 3.11 and return later if necessary.

---

[1] Classical indeed, as studies on conics date back to Menaechmus (380–320 BC) (Thomas, 1939).

Aircraft Aerodynamic Design: Geometry and Optimization, First Edition. András Sóbester and Alexander I J Forrester.
© 2015 John Wiley & Sons, Ltd. Published 2015 by John Wiley & Sons, Ltd.

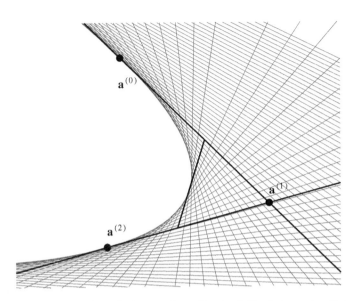

Figure 3.1   A conic as the intersections of projectivities between two lines.

3.1.1   Projective Geometry Construction of Conics

Given two points $\mathbf{a}^{(1)} = [a_1^{(1)}, a_2^{(1)}, a_3^{(1)}]'$ and $\mathbf{a}^{(2)} = [a_1^{(2)}, a_2^{(2)}, a_3^{(2)}]'$, a line through these points is defined as

$$\mathbf{A} = [A_1, A_2, A_3] = \mathbf{a}^{(1)} \times \mathbf{a}^{(2)}. \tag{3.1}$$

We view projective geometry through projections onto lines, planes and surfaces. Projecting a point onto a plane yields its image, but any other point along that projection will yield the same image; that is, points along a line are considered to be identical (just as a set of points in space aligned with the eye of an observer are indistinguishable to that observer). Note that in Equation 3.1, multiplying the points by a scalar simply shifts the points along line $\mathbf{A}$. Likewise, a line projected onto a plane is considered as identical to all other lines along that projection.

Projecting a line onto another line is a mapping called a *perspectivity*. Given lines $\mathbf{A}$ and $\mathbf{B}$ (Figure 3.2), the image of the point $\mathbf{x}$ on $\mathbf{B}$ can be found by projecting through a point $\mathbf{p}$ (the centre of the perspectivity, which is not on $\mathbf{A}$ or $\mathbf{B}$):

$$\hat{\mathbf{x}} = [\mathbf{p} \times \mathbf{x}] \times \mathbf{B}. \tag{3.2}$$

A conic is a collection of projectivities (see Figure 3.1). A projectivity from line $\mathbf{A}$ to $\mathbf{B}$ can be found from two perspectivities with the construction shown in Figure 3.3. Given points $\mathbf{a}^{(1)}$, $\mathbf{a}^{(2)}$, $\mathbf{a}^{(3)}$ and their *images* $\mathbf{b}^{(1)}$, $\mathbf{b}^{(2)}$, $\mathbf{b}^{(3)}$ (points on $\mathbf{B}$ that the projectivity we are finding would yield), we can find the intersections:

$$
\begin{aligned}
\mathbf{p}^{(1)} &= (\mathbf{a}^{(2)} \times \mathbf{b}^{(3)}) \times (\mathbf{a}^{(3)} \times \mathbf{b}^{(2)}), \\
\mathbf{p}^{(2)} &= (\mathbf{a}^{(1)} \times \mathbf{b}^{(3)}) \times (\mathbf{a}^{(3)} \times \mathbf{b}^{(1)}), \\
\mathbf{p}^{(3)} &= (\mathbf{a}^{(1)} \times \mathbf{b}^{(2)}) \times (\mathbf{a}^{(2)} \times \mathbf{b}^{(1)}).
\end{aligned}
\tag{3.3}
$$

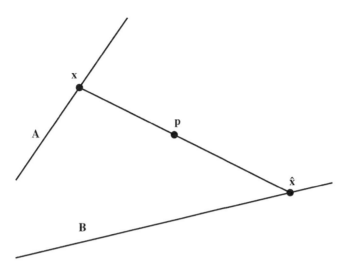

Figure 3.2   A perspectivity of line **A** onto **B**.

These points are collinear and lie on the *Pappus line*, **P**. A perspectivity from **A** to **P** with centre $\mathbf{b}^{(1)}$ maps **x** on **A** onto **q** on **P**. A second perspectivity from **P** to **B** with centre $\mathbf{a}^{(1)}$ maps **q** on **P** onto $\hat{\mathbf{x}}$ on **B**.

3.1.2   *Parametric Bernstein Conic*

We have so far performed a number of, perhaps seemingly abstract, projective geometry constructions, but what we really require is the means to define a parametric conic; that is, to find points on it.

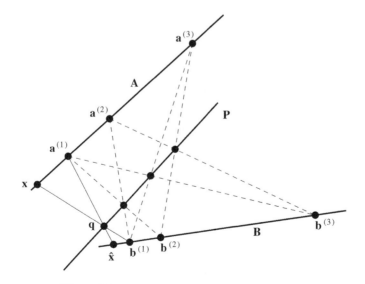

Figure 3.3   A projectivity from two perspectivities.

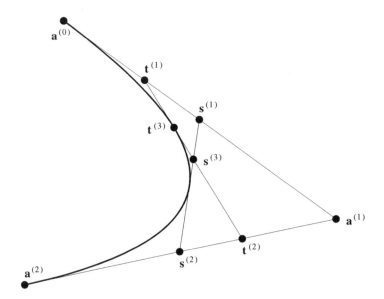

Figure 3.4   Construction of a Bernstein conic.

We see from Figure 3.3 that to define a projectivity, and so a conic, we require three pre-image and image point pairs. Figure 3.4 is a reproduction of Figure 3.1, showing these three pairs: $\mathbf{a}^{(0)} \rightarrow \mathbf{a}^{(1)}$, $\mathbf{a}^{(1)} \rightarrow \mathbf{a}^{(2)}$ and $\mathbf{s}^{(1)} \rightarrow \mathbf{s}^{(2)}$. Point $\mathbf{a}^{(1)}$ is the intersection of $\mathbf{A}$ and $\mathbf{B}$, and $\mathbf{S} = \mathbf{s}^{(1)} \times \mathbf{s}^{(2)}$ is the *shoulder tangent*. We will position this shoulder tangent at

$$\mathbf{s}^{(1)} = \frac{1}{2}\mathbf{a}^{(0)} + \frac{1}{2}\mathbf{a}^{(1)}$$

and

$$\mathbf{s}^{(2)} = \frac{1}{2}\mathbf{a}^{(1)} + \frac{1}{2}\mathbf{a}^{(2)}.$$

A point on the conic C in Figure 3.4 shall be found as the intersection of a tangent $\mathbf{T}(u)$ to the conic (u is a parameter that we vary so that $\mathbf{t}^{(3)}$ (the intersection of $\mathbf{T}$ with the conic) traces out; that is, moves along the along the conic).[2] The intersection of $\mathbf{T}$ with $\mathbf{A}$ is

$$\mathbf{t}^{(1)} = (1 - u)\mathbf{a}^{(0)} + u\mathbf{a}^{(1)}, \tag{3.4}$$

and with $\mathbf{B}$ is

$$\mathbf{t}^{(2)} = (1 - u)\mathbf{a}^{(1)} + u\mathbf{a}^{(2)}. \tag{3.5}$$

---

[2] This is a 'parameter' in the strictly mathematical sense of the word, not a design variable – we discussed the distinction in Section 2.2.

This intuitive result is derived in Farin (1999) using the four tangent theorem (which is also employed below). Thus, as u varies from 0 to 1, the intersection traces out the conic from $\mathbf{a}^{(0)}$ to $\mathbf{a}^{(2)}$ (the whole conic is traced out by $-\infty < u < \infty$).

Looking at Figure 3.4, by bringing $\mathbf{t}^{(1)}$ to $\mathbf{a}^{(1)}$ (i.e. setting $u = 0$) we see that the intersection of $\mathbf{S}$ and $\mathbf{T}$ is $\mathbf{s}^{(3)} = \frac{1}{2}\mathbf{a}^{(1)} + \frac{1}{2}\mathbf{c}$.

Four tangents to a conic (e.g. $\mathbf{A}$, $\mathbf{B}$, $\mathbf{S}$ and $\mathbf{T}$) each have an intersection with the conic and three intersections with the other three tangents. The four tangent theorem states that the cross-ratio (cr) of these four intersections equals the same constant for all four tangents. Thus:

$$\mathrm{cr}(\mathbf{a}^{(0)}, \mathbf{t}^{(1)}, \mathbf{s}^{(1)}, \mathbf{a}^{(1)}) = \mathrm{cr}(\mathbf{t}^{(1)}, \mathbf{t}^{(3)}, \mathbf{s}^{(3)}, \mathbf{t}^{(2)}). \tag{3.6}$$

Given four points on a line, such that

$$\mathbf{a}^{(3)} = \alpha_1 \mathbf{a}^{(1)} + \beta_1 \mathbf{a}^{(2)}$$

and

$$\mathbf{a}^{(4)} = \alpha_2 \mathbf{a}^{(1)} + \beta_2 \mathbf{a}^{(2)},$$

the cross-ratio of these four points is

$$\mathrm{cr}(\mathbf{a}^{(1)}, \mathbf{a}^{(3)}, \mathbf{a}^{(4)}, \mathbf{a}^{(2)}) = \frac{\beta_1}{\alpha_1}\frac{\alpha_2}{\beta_2}. \tag{3.7}$$

Using Equations 3.7 and 3.6,

$$\mathrm{cr}(\mathbf{t}^{(1)}, \mathbf{t}^{(3)}, \mathbf{s}^{(3)}, \mathbf{t}^{(2)}) = \frac{1/2}{1/2}\frac{u}{1-u}. \tag{3.8}$$

And so:

$$\mathbf{t}^{(3)}(u) = (1-t)\mathbf{t}^{(1)}(u) + t\mathbf{t}^{(2)}(u). \tag{3.9}$$

Substituting Equations 3.4 and 3.5 into Equation 3.9, we find that the point on the conic, which we will now call $C(t)$, is a quadratic with coefficients based on our original points in Figure 3.4:

$$C(u) = \mathbf{t}^{(3)}(u) = (1-u)^2\mathbf{a}^{(0)} + 2u(1-u)\mathbf{a}^{(1)} + u^2\mathbf{a}^{(2)}. \tag{3.10}$$

With the quadratic Bernstein basis polynomials defined as

$$b_{i,2}(u) = \binom{2}{i} u^i (1-u)^{2-i}, \quad i = 0, 1, 2,$$ (3.11)

Equation 3.10 can be rewritten as a projective quadratic Bézier curve:

$$B(u) = C(u) = \mathbf{t}^{(3)}(u) = \sum_{i=0}^{2} \mathbf{a}^{(i)} b_{i,2}(u).$$ (3.12)

We see that the Bézier curve is a weighted sum of the *control points*, $\mathbf{a}^{(i)}$. The influence of these control points, which is determined by their weighting $b_{i,2}(u)$, should be greatest when the curve passes closest to them, and this influence should diminish as the curve moves away. For example, $b_{i,2}(u)$ should reach a maximum when $u = 0.5$ and $b_{i,2}(u) \to 0$ as $t \to \pm\infty$. Figure 3.5 shows the Bernstein polynomial weightings (Equation 3.11) used by the Bézier curve.

Clearly, it is possible to add further control points to the sum in Equation 3.12, weighted using higher degree Bernstein polynomials:

$$b_{i,n}(u) = \binom{n}{i} u^i (1-u)^{n-i}, \quad \text{where} \quad \binom{n}{i} = \frac{n!}{i!(n-i)!}.$$ (3.13)

The third-degree Bernstein polynomials are shown in Figure 3.6.

Figure 3.7 demonstrates this by inserting a control point in to the definition of the quadratic Bézier curve in Figure 3.4 to produce a cubic Bézier curve. The MATLAB® code to produce

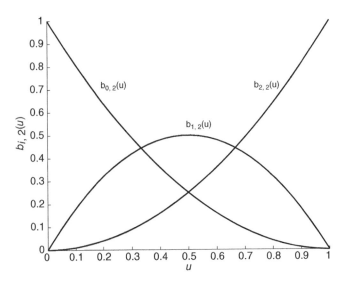

Figure 3.5 The Bernstein polynomials (Equation 3.11), displaying the intuitive property of Bézier curve control point weighting falling away from control points.

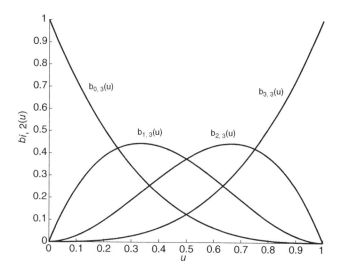

Figure 3.6   The third degree Bernstein polynomials (Equation 3.13).

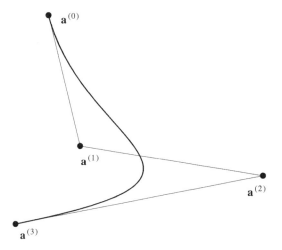

Figure 3.7   A cubic Bézier curve produced by inserting an additional control point to the quadratic Bézier curve in Figure 3.4.

this curve is in Listing 3.1 and the OpenNURBS/Rhino-Python code is in Listing 3.2. Listing 3.3 is the OpenNURBS/Rhino-Python code to produce a generic Bézier curve with interactive entry of control points.[3]

---

[3] OpenNURBS/Rhino-Python jumps from `AddCurve` to `AddNurbsCurve`, without the intermediate curves we will consider in Sections 3.1.3, 3.2 and 3.4. We will return to OpenNURBS/Rhino-Python implementations in our discussion of NURBS in Section 3.6, and there show how the curves in the preceding sections can be constructed with this more flexible method.

Listing 3.1 MATLAB® code to produce the cubic Bézier curve in Figure 3.7.

```
1   % four control points
2   a0=[1 10 1]';
3   a1=[2 5 1]';
4   a2=[7.7 3.9 1]';
5   a3=[0 2 1]';
6   % calculate projective cubic bezier for t[0,1]
7   % and project onto z=1 plane
8   t=0:0.01:1;
9   for i=1:101
10      cubicBezier(i,:)=(1-t(i))^3*a0+...
11                          3*t(i)*(1-t(i))^2*a1+...
12                          3*t(i)^2*(1-t(i))*a2+...
13                          t(i)^3*a3;
14      cubicBezier(i,:)=cubicBezier(i,:)./cubicBezier(i,3);
15  end
16  % plot
17  plot(cubicBezier(:,1),cubicBezier(:,2),'k','LineWidth',2)
```

Listing 3.2 OpenNURBS/Rhino-Python code to produce the cubic Bézier curve in Figure 3.7.

```
1   import rhinoscriptsyntax as rs
2   # four control points
3   a0=(1, 10, 1)
4   a1=(2, 5, 1)
5   a2=(7.7, 3.9, 1)
6   a3=(0, 2, 1)
7   # call AddCurve with control points and degree 3
8   rs.AddCurve((a0,a1,a2,a3),3)
```

Listing 3.3 OpenNURBS/Rhino-Python code to produce a generic Bézier curve with interactive entry of control points.

```
1   import rhinoscriptsyntax as rs
2   # user-defined control points
3   points = rs.GetPoints(True, message1="Pick curve point")
4   # degree is number of control points -1
5   degree = len(points)-1
6   # call AddCurve with user points and degree
7   if points: rs.AddCurve(points,degree)
```

A Bézier curve can be expressed in matrix form as

$$B_i(t) = (u^d, u^{d-1}, \ldots, u, 1)\mathbf{M}^{(d)} \begin{pmatrix} \mathbf{a}^{(i-1)} \\ \mathbf{a}^{(i)} \\ \vdots \\ \mathbf{a}^{(i+d-1)} \end{pmatrix}, \tag{3.14}$$

where elements of the basis matrix $\mathbf{M}$ are given by

$$m_{i,j}^{(d)} = \binom{d}{j-1}\binom{d-j+1}{d-j-i+2}(-1)^{(d-j-i+2)} \tag{3.15}$$

$$\text{for } i = 1, \ldots, d, j = 1, \ldots, d - (i-1). \tag{3.16}$$

The first four $\mathbf{M}$ matrices are

$$\mathbf{M}^{(d=1)} = \begin{pmatrix} -1 & 1 \\ 1 & 0 \end{pmatrix},$$

$$\mathbf{M}^{(d=2)} = \begin{pmatrix} 1 & -2 & 1 \\ -2 & 2 & 0 \\ 1 & 0 & 0 \end{pmatrix},$$

$$\mathbf{M}^{(d=3)} = \begin{pmatrix} -1 & 3 & -3 & 1 \\ 3 & -6 & 3 & 0 \\ -3 & 3 & 0 & 0 \\ 1 & 0 & 0 & 0 \end{pmatrix},$$

$$\mathbf{M}^{(d=4)} = \begin{pmatrix} 1 & -4 & 6 & -4 & 1 \\ -4 & 12 & -12 & 4 & 0 \\ 6 & -6 & -6 & 0 & 0 \\ -4 & 4 & 0 & 0 & 0 \\ 1 & 0 & 0 & 0 & 0 \end{pmatrix}.$$

3.1.3 Rational Conics and Bézier Curves

While the use of projective geometry to define conics is elegant and intuitive, it is unsuitable for design purposes. We can project onto the plane $z = 1$ (an affine plane) simply by dividing the projective points through by the z-component (the figures in the previous sections are in this plane):

$$\mathbf{a} = \begin{pmatrix} a_x \\ a_y \end{pmatrix} = \frac{\mathbf{a}^P}{a_z^P} = \begin{pmatrix} a_x^P/a_z^P \\ a_y^P/a_z^P \\ a_z^P/a_z^P \end{pmatrix},$$

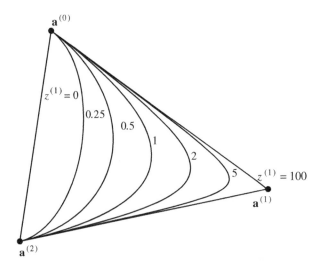

Figure 3.8 Varying weights for a rational quadratic conic.

where here we use $\mathbb{P}$ to denote a point in projective space. Equation 3.10 can now be written in *rational* form as

$$C(u) = \frac{(1-u)^2 z^{(0)} \mathbf{a}^{(0)} + 2u(1-u)z^{(1)}\mathbf{a}^{(1)} + u^2 z^{(2)}\mathbf{a}^{(2)}}{(1-u)^2 z^{(0)} + 2u(1-u)z^{(1)} + u^2 z^{(2)}}, \tag{3.17}$$

(and also for higher degree conics) where, for example, $\mathbf{a}_z^{\mathbb{P}(0)}$ is written simply as $z^{(0)}$. These $z^{(i)}$ variables are known as *weights* and the $\mathbf{a}^{(i)}$ variables are the control points.

In a similar way, Equation 3.12 can now be cast as a rational Bézier curve; Equation 3.10 can be rewritten as a projective quadratic Bézier curve:

$$B(u) = \frac{\sum_{i=0}^{n} w_i \mathbf{a}^{(i)} b_{i,n}(u)}{\sum_{i=0}^{n} w_i b_{i,n}(u)}, \tag{3.18}$$

where w_i are the weights. Figure 3.4 has all weights set at unity. Figure 3.8 shows the effect of varying $z^{(1)}$, with $z^{(0)}, z^{(2)} = 1$. The effect is intuitive; for $z^{(1)} = 0$, C becomes a straight line from $\mathbf{a}^{(0)}$ to $\mathbf{a}^{(2)}$, and as $z^{(1)} \to \infty$, $C(t)$ tends to the line $\mathbf{a}^{(0)}, \mathbf{a}^{(1)}, \mathbf{a}^{(2)}$.

For the remainder of this chapter we will consider rational curves defined by x, y coordinates of points and weightings.

3.1.4 Properties of Bézier Curves

The above derivation of a conic, and so a Bézier curve, has led to an intuitive form of curve, which can be 'pushed and pulled' around by control points. Salomon (2006) lists a number of properties of Bézier curves (and discusses them in more detail than we shall do here) that help in understanding their usefulness in geometry definition and manipulation:

- the weights $b_{i,n}(u)$ add up to 1 (they are *barycentric*);
- the curve passes through the two endpoints $\mathbf{a}^{(0)}$ and $\mathbf{a}^{(n)}$;
- the curve is symmetric with respect to the numbering of control points – that is, the same result will be obtained by setting $\mathbf{a}^{(0)} = \mathbf{a}^{(n)}$, $\mathbf{a}^{(1)} = \mathbf{a}^{(n-1)}$, ..., $\mathbf{a}^{(n)} = \mathbf{a}^{(0)}$;
- the first derivative is simple to obtain as

$$\dot{\mathbf{a}}(u) = \Delta\mathbf{a}^{(i)} b_{i,n-1}(u), \quad \text{where} \quad \Delta\mathbf{a}^{(i)} = \mathbf{a}^{(i+1)} - \mathbf{a}^{(i)}; \tag{3.19}$$

- the weight functions $b_{i,n}$ have a maximum at $u = i/n$ – that is, the influence of the control points is spaced evenly with respect to t;
- the two end tangents, derived from Equation 3.19 as $\dot{\mathbf{a}}^{(0)} = n(\mathbf{a}^{(1)} - \mathbf{a}^{(0)})$ and $\dot{\mathbf{a}}(1) = n(\mathbf{a}^{(n)} - \mathbf{a}^{(n-1)})$, are easily controlled by moving $\mathbf{a}^{(1)}$ and $\mathbf{a}^{(n-1)}$ respectively;
- the Bézier curve is controlled globally by editing control point(s) and weighting(s) – that is, changing one control point will affect the entire curve (note how in Figures 3.5 and 3.6 the weighting of each point has an influence over all values of u), with the effect greatest close to the control point;
- the curve is contained within the control polygon (within its *convex hull*), as seen in Figure 3.7, where the curve lies within the triangle formed by $\mathbf{a}^{(0)}, \mathbf{a}^{(1)}, \mathbf{a}^{(2)}$ and then that formed by $\mathbf{a}^{(1)}, \mathbf{a}^{(2)}, \mathbf{a}^{(3)}$; and
- the curve is invariant to affine transformations (e.g. can be shifted up/down, left/right, rotated, reflected), but is not invariant under projections.

The following sections will show how the curve can be controlled by varying the control point weights and linking together conics to arrive at the well-known NURBS definition.

3.2 Bézier Splines

A series of conics or Bézier curves can be concatenated to produce a more complex curve. The process is quite straightforward. For two successive curves defined by control points $\mathbf{p}^{(i)}$, $i = 1, 2, \ldots, n$ and $\mathbf{q}^{(i)}$, $i = 1, 2, \ldots, m$ to meet, the end and start points must coincide; that is, $\mathbf{p}^{(n)} = \mathbf{q}^{(0)}$. For a smooth connection, $\dot{\mathbf{p}}^{(n)} = \dot{\mathbf{q}}^{(0)}$. Thus, from Equation 3.19 and the end tangent note above,

$$\mathbf{p}^{(n)} = \mathbf{q}^{(0)} = \frac{m}{m+n}\mathbf{q}^{(1)} + \frac{n}{m+n}\mathbf{p}^{(n-1)}. \tag{3.20}$$

Note that we do not need to specify the first and nth points, as these are found from the points either side of the connection. Figure 3.9 shows the concatenation of two quadratic Bézier curves to form a Bézier spline, while Figure 3.10 shows two cubic Bézier curves joined into a Bézier spline representation of an aerofoil. The MATLAB® code to produce the spline in this figure is included in Listing 3.4 (the code is rather long-winded, as we have calculated the cubic Bézier curves explicitly, rather than using the generic formulation for the Bernstein polynomials in Equation 3.13, as we have in the toolkit accompanying this book). By changing

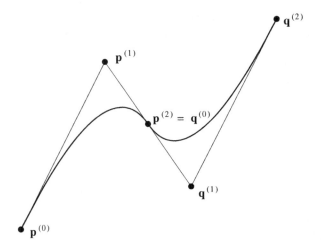

Figure 3.9   An example of a Bézier spline formed by joining two quadratic Bézier curves.

six parameters – $\mathbf{p}_y^{(2)}$, $\mathbf{q}_y^{(1)}$, $\mathbf{p}_{x\&y}^{(1)}$, $\mathbf{q}_{x\&y}^{(2)}$ – a good degree of shape control is possible. Further control can be had by manipulating the weights $z^{(i)}$, and we will consider this later.

3.3   Ferguson's Spline

Introduced by the Boeing Company's James Ferguson (1964), this is a formulation with two distinguishing features. First, as we are about to see, it is, mathematically, very simple. Second, its behaviour is beautifully intuitive. From an optimization perspective, this is a rather handy trait when we are attempting to establish suitable bounds on the design variables.

Fundamentally, a Ferguson spline is a curve $\mathbf{r}(u)$ (with the parameter $u \in [0, 1]$), connecting two points $\mathbf{r}(0) = \mathbf{A}$ and $\mathbf{r}(0) = \mathbf{B}$ in such a way that its tangent has a given value at these end points: $d\mathbf{r}/du|_{u=0} = \mathbf{T}_A$ and $d\mathbf{r}/du|_{u=1} = \mathbf{T}_B$ – see the illustration in Figure 3.11.

We define the curve as the cubic polynomial

$$\mathbf{r}(u) = \sum_{i=0}^{3} \mathbf{a}_i u^i, \quad u \in [0, 1]. \tag{3.21}$$

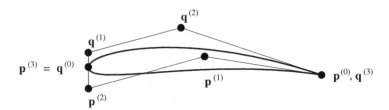

Figure 3.10   An example of a Bézier spline representation of an aerofoil, formed by joining two cubic Bézier curves.

Listing 3.4 MATLAB® code for generating the Bézier spline aerofoil in Figure 3.10.

```matlab
% control points
p0=[1 0]';
p1=[0.5 0.08]';
p2=[0 -0.05]';
q1=[0 0.1]';
q2=[0.4 0.2]';
q3=[1 0]';

% weights
zp=[1 1 1 1];

zq=[1 1 1 1];

% calculate connection point
p3=(2/(2+3))*p2+(3/(2+3))*q1;
q0=p3;

figure
hold on

% calculate rational cubic Bezier for t[0,1]
t=0:0.01:1;
for i=1:101
    lower(i,:)=((1-t(i))^3*zp(1)*p0+...
        3*t(i)*(1-t(i))^2*zp(2)*p1+...
        3*t(i)^2*(1-t(i))*zp(3)*p2+...
        t(i)^3*zp(4)*p3)./...
        ((1-t(i))^3*zp(1)+...
        3*t(i)*(1-t(i))^2*zp(2)+...
        3*t(i)^2*(1-t(i))*zp(3)+...
        t(i)^3*zp(4));
    upper(i,:)=((1-t(i))^3*zq(1)*q0+...
        3*t(i)*(1-t(i))^2*zq(2)*q1+...
        3*t(i)^2*(1-t(i))*zq(3)*q2+...
        t(i)^3*zq(4)*q3)./...
        ((1-t(i))^3*zq(1)+...
        3*t(i)*(1-t(i))^2*zq(2)+...
        3*t(i)^2*(1-t(i))*zq(3)+...
        t(i)^3*zq(4));
end
% plot
plot(lower(:,1),lower(:,2),'k','LineWidth',2)
plot(upper(:,1),upper(:,2),'k','LineWidth',2)
```

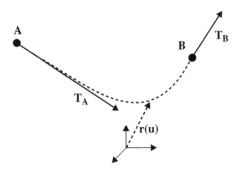

Figure 3.11   Ferguson spline and its boundary conditions.

We find the four sets of numbers required to define the curve by setting the endpoint conditions:

$$
\begin{aligned}
\mathbf{A} &= \mathbf{a}_0 \\
\mathbf{B} &= \mathbf{a}_0 + \mathbf{a}_1 + \mathbf{a}_2 + \mathbf{a}_3 \\
\mathbf{T_A} &= \mathbf{a}_1 \\
\mathbf{T_B} &= \mathbf{a}_1 + 2\mathbf{a}_2 + 3\mathbf{a}_3.
\end{aligned}
\tag{3.22}
$$

Rearranging in terms of the vectors:

$$
\begin{aligned}
\mathbf{a}_0 &= \mathbf{A} \\
\mathbf{a}_1 &= \mathbf{T_A} \\
\mathbf{a}_2 &= 3[\mathbf{B} - \mathbf{A}] - 2\mathbf{T_A} - \mathbf{T_A} \\
\mathbf{a}_3 &= 2[\mathbf{A} - \mathbf{B}] + \mathbf{T_A} + \mathbf{T_B}
\end{aligned}
\tag{3.23}
$$

Substituting back into (3.21) we obtain

$$
\mathbf{r}(u) = \mathbf{A}(1 - 3u^2 + 2u^3) + \mathbf{B}(3u^2 - 2u^3) + \mathbf{T_A}(u - 2u^2 + u^3) + \mathbf{T_B}(-u^2 + u^3), \tag{3.24}
$$

or in matrix form:

$$
\mathbf{r}(u) = \begin{bmatrix} 1 & u & u^2 & u^3 \end{bmatrix}
\begin{bmatrix}
1 & 0 & 0 & 0 \\
0 & 0 & 1 & 0 \\
-3 & 3 & -2 & -1 \\
2 & -2 & 1 & 1
\end{bmatrix}
\begin{bmatrix}
\mathbf{A} \\
\mathbf{B} \\
\mathbf{T_A} \\
\mathbf{T_B}
\end{bmatrix}.
\tag{3.25}
$$

$\mathbf{r}(u)$ is, essentially, a Hermitian interpolant (i.e. an interpolant with specified gradients at the interpolation points) and the bracketed factors in (3.24) can be viewed as its basis functions – Figure 3.12 illustrates their effect on the shape of the interpolant.

There is a clear similarity between the Ferguson spline and a Bézier curve. Indeed, we can view the Ferguson spline as a reparameterized Bézier curve. While the end tangents of a Bézier

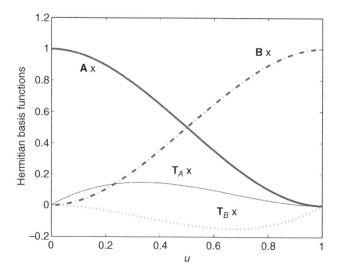

Figure 3.12   The four basis functions of equation (3.24) shown alongside their respective multipliers.

curve are controlled by shifting control points, the underlying control points of a Ferguson spline are controlled by modifying end tangents directly. Both are a weighted sum of basis functions, with Figure 3.6 showing the basis functions for the Bézier curve and Figure 3.12 those for the Ferguson spline. Both have the same characteristics of global control and the end points being the sole contributor to the curve at its ends.

In Section 3.1.4 we covered the key benefits of Bézier curves, noting that the end tangents can easily be modified by moving the second and $(n-1)$ control points. The Ferguson spline takes things a step further in terms of 'user friendliness' by allowing us to define explicitly the end tangents. By user friendliness we are thinking in terms of the discussion of what makes a good parametric geometry in Section 2.3.

The simplicity of the Ferguson splines is illustrated by the conciseness of their MATLAB® implementation (Listing 3.5, to be found as `hermite.m` as part of the toolset accompanying this book).

From a geometry parameterization perspective, the Ferguson scheme gives us a tangent vector – direction and magnitude – to play with. The greater the magnitude, the 'stiffer' the curve will become in the neighbourhood of the corresponding start/end point. This is illustrated in Figure 3.13, obtained by running the the small sample code (`hermite_example.m`) shown in Listing 3.6 (which calls a slightly more complete version of the function of Listing 3.5, one that generates a plot of the curve if the variable `PlotReq` is set to a nonzero number).

By scaling the vector of the starting point with different values (maintaining the direction), the effect of the magnitude becomes clear – in Figure 3.13, the lower curves correspond to the shorter vectors (visualizing the tangent at the point $(1, 2)$). The reader may wish to conduct similar variable sweeps by editing `hermite_example.m` to, say, vary the directions of the two vectors too.

In Section 7.2 we shall also illustrate the power of this simple formulation via a parametric aerofoil section geometry based on a pair of Ferguson splines, one for each surface of the aerofoil.

Listing 3.5 MATLAB® code for generating a Hermite interpolant – an implementation of (3.25).

```matlab
 1  function r = hermite(r0, r1, dr_by_du0, dr_by_du1, divisions)
 2
 3  % Hermite basis function matrix
 4  C = [1 0 0 0; 0 0 1 0; -3 3 -2 -1; 2 -2 1 1];
 5
 6  S = [          r0(:)';
 7                 r1(:)';
 8         dr_by_du0(:)';
 9         dr_by_du1(:)' ];
10
11  u = (0:1/divisions:1);
12
13  for i=1:length(u)
14      U = [1 u(i) u(i)^2 u(i)^3];
15      r(i,:) = (U*C)*S;
16  end
```

Listing 3.6 Figure 3.13 was generated using this short MATLAB® script.

```matlab
 1  r0 = [ 1 2 ];
 2  r1 = [ 4 1.5 ];
 3  dr_by_du1 = [ 1 0 ];
 4  divisions = 50;
 5  PlotReq = 1;
 6
 7  for i = (0.5:1:4.5)
 8      dr_by_du0 = [ 1   0.1 ]*i;
 9      r = hermite(r0, r1, dr_by_du0, dr_by_du1, divisions, PlotReq);
10  end
```

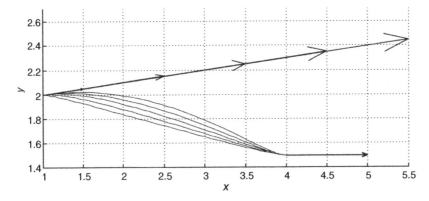

Figure 3.13 The effect of varying the tangent magnitude at point $(1,2)$ – the higher curves correspond to the longer tangent vectors.

3.4 B-Splines

B-splines, where the 'B' stands for 'basis', have a number of key characteristics that may make them preferable to Bézier splines in many applications:

- The degree of a B-spline is not determined by the number of control points.
- Local control of the spline is possible (unlike Bézier splines, where moving one control point affects the whole spline).
- The degree of continuity between segments can be specified.

This is not an exhaustive list, but encapsulates the key differences from a practical perspective.

Local control of the spline means that moving or changing the weight of a control point will only affect the spline in the vicinity of this point. The extent of this vicinity is determined by the degree of the B-spline. In Figure 3.14 we see that moving control point $\mathbf{a}^{(0)}$ only affects a degree-2 spline up until midway between $\mathbf{a}^{(1)}$ and $\mathbf{a}^{(3)}$ (up until the end of the first segment).

A B-spline is made of $(n + 1) - d$ segments, which we shall call $B_0, B_1, \ldots, B_{(n-d)}$, where $n + 1$ is the number of control points (remember these are numbered from zero) and d is the degree of each segment (the degree of the B-spline). The segments start and end at joints $\mathbf{k}^{(0)}, \mathbf{k}^{(1)}, \ldots, \mathbf{k}^{((n+1)-d)}$. Note that, unlike a Bézier spline, the ends of the B-spline $\mathbf{k}^{(0)}$ and $\mathbf{k}^{((n+1)-d)}$ do not coincide with the first and last control points. The exception to this is if there are multiple control points: a degree-d B-spline passes through a control point repeated d times. For example, if $\mathbf{a}^{(0)} = \mathbf{a}^{(1)}$, a degree-2 B-spline will start at the first control point (the first segment will be a straight line towards $\mathbf{a}^{(2)}$).

We will start by defining a uniform B-spline using control points and then go on to interpolating B-splines; that is, defined by the joint points, which may be more practical in some situations.

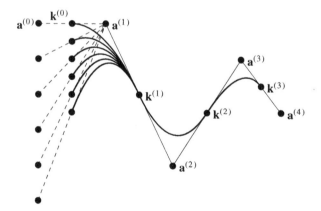

Figure 3.14 A quadratic rational B-spline showing the vicinity of local control as control point $\mathbf{a}^{(0)}$ is shifted.

Expressed in the same way as Equation 3.14, the ith segment of a degree-d uniform B-spline can be calculated as

$$B_i(t) = (u^d, u^{d-1}, \ldots, u, 1)\mathbf{M}^{(d)} \begin{pmatrix} \mathbf{a}^{(i-1)} \\ \mathbf{a}^{(i)} \\ \vdots \\ \mathbf{a}^{(i+d-1)} \end{pmatrix}, \tag{3.26}$$

where elements of the basis matrix $\mathbf{M}$ are given by

$$m_{i,j}^{(d)} = \frac{1}{d!}\binom{n}{i}\sum_{k=j}^{d}(d-k)^i(-1)^{d-j}\binom{d+1}{k-j}. \tag{3.27}$$

The first four $\mathbf{M}$ matrices are

$$\mathbf{M}^{(d=1)} = \frac{1}{1!}\begin{pmatrix} -1 & 1 \\ 1 & 0 \end{pmatrix},$$

$$\mathbf{M}^{(d=2)} = \frac{1}{2!}\begin{pmatrix} 1 & -2 & 1 \\ -2 & 2 & 0 \\ 1 & 1 & 0 \end{pmatrix},$$

$$\mathbf{M}^{(d=3)} = \frac{1}{3!}\begin{pmatrix} -1 & 3 & -3 & 1 \\ 3 & -6 & 3 & 0 \\ -3 & 0 & 3 & 0 \\ 1 & 4 & 1 & 0 \end{pmatrix},$$

$$\mathbf{M}^{(d=4)} = \frac{1}{4!}\begin{pmatrix} 1 & -4 & 6 & -4 & 1 \\ -4 & 12 & -12 & 4 & 0 \\ 6 & -6 & -6 & 6 & 0 \\ -4 & 12 & 12 & 4 & 0 \\ 1 & 11 & 11 & 1 & 0 \end{pmatrix}.$$

Varying t from zero to one, Equation 3.26 traces out the ith segment of the B-spline between joints $\mathbf{k}^{(i)}$ and $\mathbf{k}^{(i+1)}$. These joints can be calculated as

$$\mathbf{k}^{(i)} = \frac{1}{d!}(\mathbf{a}^{(i)}\mathbf{a}^{(1)}\ldots\mathbf{a}^{(i+d-1)})\begin{pmatrix} c_0^{(d)} \\ c_1^{(d)} \\ \vdots \\ c_{d-1}^{(d)} \end{pmatrix}, \tag{3.28}$$

where

$$c_j^{(d)} = \sum_{k=j}^{d} (d-k)^d (-1)^{k-j} \binom{d+1}{k-j}$$

(3.29)

$(c_{0,1,\ldots,d}^{(d)}$ is also the first d elements of the last row of the $\mathbf{M}^{(d)}$ matrix).

Having found the joint points, given the control points, we can now find the control points (and so the B-spline) given the joint points. We actually need a little more information. A B-spline with $n+1$ control points has $(n+1) - d + 1$ joints; so, along with the $(n+1) - d + 1$ equations based on these joints, we need $d-1$ further equations to calculate the $n+1$ control points. Specifying tangents gives us the remaining expressions. For a quadratic B-spline we only need to specify one tangent (e.g. the start tangent), resulting in the following system of equations:

$$\frac{1}{2!}\begin{pmatrix} -2 & 2 & 0 & \cdots & 0 & 0 \\ c_0^{(2)} & c_1^{(2)} & 0 & \cdots & 0 & 0 \\ 0 & c_0^{(2)} & c_1^{(2)} & \cdots & 0 & 0 \\ \vdots & \vdots & \vdots & \ddots & \ddots & \vdots \\ 0 & 0 & 0 & \cdots & c_0^{(2)} & c_1^{(2)} \end{pmatrix} \begin{pmatrix} \mathbf{a}^{(0)} \\ \mathbf{a}^{(1)} \\ \mathbf{a}^{(2)} \\ \vdots \\ \mathbf{a}^{(n)} \end{pmatrix} = \begin{pmatrix} \mathbf{t}^{(0)} \\ \mathbf{k}^{(0)} \\ \mathbf{k}^{(1)} \\ \vdots \\ \mathbf{k}^{(n+1-d)} \end{pmatrix},$$

(3.30)

which can easily be solved to find $\mathbf{a}^{(0,\ldots,n)}$.

We can use the above to define a B-spline through a set of joints, yielding a versatile curve with intuitive, local control via the calculated control points. However, specifying a mix of joints and tangents may be undesirable when compared with the convenient method of defining a Bézier curve with start, end and intermediate control points. An answer lies in the use of a *knot vector*.

3.5   Knots

Each segment of a B-spline is defined over an interval where t varies $\in [0, 1]$ and the curve is defined, in this interval by $d+1$ control points. For example, from Equation 3.26 we see the first segment of a degree-3 B-spline is defined by $\mathbf{a}^{(0)}, \mathbf{a}^{(1)}, \ldots, \mathbf{a}^{(3)}$ multiplied by some weightings (basis functions; the 'B' in B-spline). These basis functions overlap from segment to segment. Figure 3.15 shows how four basis functions are 'in play' in each interval for a degree-3 B-spline. Referring back to Equation 3.26, in each interval, $\mathbf{a}^{(i-1)}$ is multiplied by the basis function decreasing from $\frac{1}{6}$, $\mathbf{a}^{(i)}$ by the one decreasing from $\frac{2}{3}$, $\mathbf{a}^{(i+1)}$ by the one increasing from $\frac{1}{6}$, and $\mathbf{a}^{(i+2)}$ by the one increasing from 0. Note how similar these basis functions are to Bézier curve weightings, except that, since they overlap, the curve does not pass through any control points.

The start and end of the intervals can be defined as a sequence of knots in a nondecreasing vector with unit, uniform spacing (e.g. $\mathbf{k} = [-10123]$, as in Figure 3.15). The parameter u varies $\in [0, 1]$ between each of these knots. In practice, we work with the knot vector when defining the B-spline and convert to $u \in [0, 1]$ 'in the maths'. The power of this knot-vector

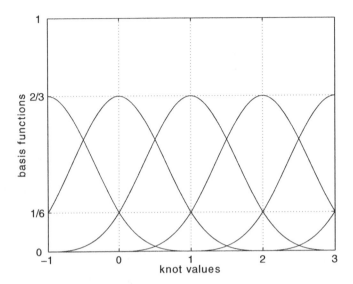

Figure 3.15   Overlapping degree-3 B-spline basis functions.

definition is when we make it nonuniform, with repeated values and nonunit intervals. This produces a nonuniform B-spline, and we only need to make it rational to create a NURBS. All we need to do is multiply by control point weights and divide through by the sum of the product of these weights and the basis function (in the same way as in Equation 3.18) and we have a NURBS.

3.6   Nonuniform Rational Basis Splines

By varying the knot vector in a nonuniform manner, we can achieve an increased level of versatility and recoup some of the attractive features of Bézier curves. The first feature to note is that repeating a knot value $d + 1$ times results in the NURBS passing through a control points. The knot vector is $(n + 1) + (d + 1)$ long, so for a degree-2 NURBS with three control points there are six knots.[4] Repeating the first and last knot values three times will create a NURBS with one quadratic segment that starts and ends at the first and last control points. The basis functions for this NURBS are shown in Figure 3.16. We see that they are the same as those for a Bézier curve (Figure 3.5). A Bézier curve is indeed a special case of a NURBS from which, using the knot vector and the ability to specify the degree of the NURBS, we can add more and more complexity.

---

[4] The nomenclature and numbering of control points gets a little confusing here. Recall that n is the index of the last control point starting from zero, so there are $n + 1$ control points. The degree of each segment of the NURBS is d and the order is $d + 1$. Thus, it is perhaps easier to think of the number of knots as the number of control points plus the order of the NURBS. We use this seemingly awkward notation and numbering to fit in with most texts on the subject and computer code indexing convention (although MATLAB® indexes from 1!).

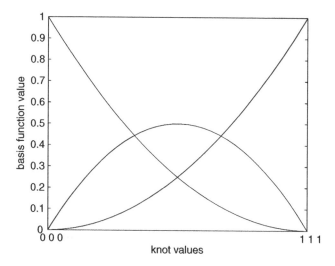

Figure 3.16 Repeated knot values in a third-degree NURBS yield the cubic Bernstein polynomials in Figure 3.6 (i.e. the NURBS is equivalent to a Bézier curve).

We can take advantage of NURBS local control to, for example, refine the shape of the Bézier curve aerofoil definition in Figure 3.10. With the influence of a control point extending over $d + 1$ knot values, adding an extra control point will not give local control in our degree-3 definition. Reducing to degree 2 *and* adding a control point gives local control, as shown in Figure 3.17, where moving a control point on the lower surface gives control over the aft portion of the aerofoil without affecting the leading edge region.

The knot vector for the lower surface NURBS is $\mathbf{k} = [0, 0, 0, 1, 2, 3, 3, 3]$ and the resulting basis functions are shown in Figure 3.18. Note how the second basis function is nonzero only for the first and second segments; that is, it has no influence on the last segment (the leading edge region of the aerofoil in Figure 3.17).

A further level of detail can be obtained by varying the spacing of the knot vector. Bringing knots together pulls the NURBS towards the control polygon: for a degree-3 NURBS, two repeated knot values matches the gradient of the NURBS at the knot to the control polygon (i.e. aligns it with a line drawn between the knots either side), three repeated knot values causes the NURBS to pass through a control point (with a gradient discontinuity) and four repeated knots causes a break in the NURBS between two control points.

Figure 3.19 shows the range of 'manoeuvres' possible with a degree-3 NURBS (omitting the effect of changing control point weights, which is similar to changing rational Bézier spline weights, as shown in Figure 3.8). Figure 3.19a is a uniform degree-3 B-spline, and the uniform weightings from the uniform knot vector are shown in the right-hand plot. Figure 3.19b has the degree increased to 7 and the first and last knot values repeated to yield a Bézier spline equivalent. In Figure 3.19c the degree is reduced back to 3 and the first and last knots repeated such that the NURBS starts and ends at the first and last control points. Figure 3.19c–f shows the effect of repeated knot values: causing tangent matching, control point interpolation and breaks in the NURBS.

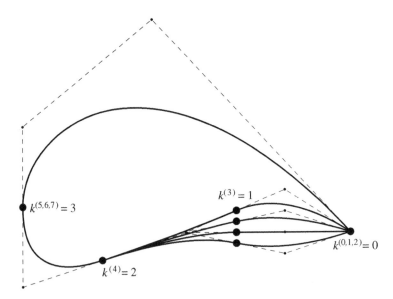

Figure 3.17 Reducing the degree of the lower surface NURBS to 3 and adding a control point gives local control towards the trailing edge, potentially giving more design capability than the Bézier curve definition in Figure 3.10 (shown here by moving the additional control point). Note that the plot has been stretched to highlight the geometry changes.

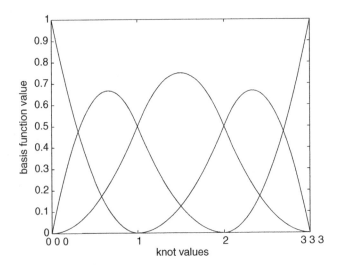

Figure 3.18 Degree 2 NURBS basis functions for knot vector $\mathbf{k} = [0,0,0,1,2,3,3,3]$, resulting in a curve starting and ending at the first and last control points. The central segment is 'uniform', but the first and last segments have basis functions distorted by the repeated knot values.

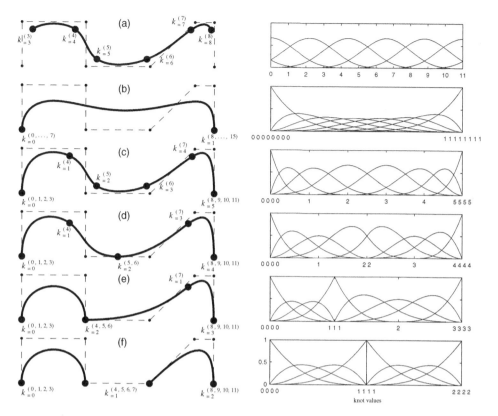

Figure 3.19 Various NURBS defined by eight control points (left) and knot vectors (right): (a) degree 3 with uniform knot vector; (b) degree 7; (c–f) degree 3. The control point locations are chosen to be similar to Salomon (2006, figure 7.19).

As a weighted sum of control points, the equation for a NURBS can be expressed in a similar way as a Bézier curve:

$$\mathcal{N}(u) = \frac{\sum_{i=0}^{n} w_i \mathbf{a}^{(i)} B_{i,d}(u)}{\sum_{i=0}^{n} w_i B_{i,d}(u)}. \tag{3.31}$$

However, since the basis functions $B_{i,d}(u)$ depend on the knot vector, the calculation of $B_{i,d}(u)$ is rather tedious and usually performed *recursively*. That is, we start by calculating $B_{0,1}(u)$ and $B_{1,1}(t)$, from which $B_{0,2}(u)$ can be calculated, and so on through to $B_{n,d}(u)$.

The basis functions for $d = 1$ are defined as

$$B_{i,1}(u) = \begin{cases} 1, & \text{if } u \in [i, i+1] \\ 0, & \text{otherwise} \end{cases}, \tag{3.32}$$

Listing 3.7   MATLAB® code to calculate NURBS basis functions using Equations 3.32 and 3.33. Note: remember that MATLAB® indexing starts at 0 and the equations start from zero.

```
 1  function B=nurbsweight(kn,degree,u)
 2  B=zeros(length(kn)-1,degree+1);
 3  for k=1:degree+1
 4      for i=1:length(kn)-k
 5          if k==1 % intitialize recursive calculation at d=1
 6              if (u>kn(i) \&\& u<=kn(i+1)) || (u==kn(1) \&\& i==degree+1)
 7                  B(i,k)=1;
 8              else
 9                  B(i,k)=0;
10              end
11          else % d>1 basis functions
12              % catch denominator=0 cases
13              if kn(i+k-1)==kn(i) \&\& kn(i+k)==kn(i+1)
14                  B(i,k)=0;
15              elseif kn(i+k)==kn(i+1)
16                  B(i,k)=((u-kn(i))/(kn(i+k-1)-kn(i)))*B(i,k-1);
17              elseif kn(i+k-1)==kn(i)
18                  B(i,k)=((kn(i+k)-u)/(kn(i+k)-kn(i+1)))*B(i+1,k-1);
19              else
20                  B(i,k)=((u-kn(i))/(kn(i+k-1)-kn(i)))*B(i,k-1)+...
21                      ((kn(i+k)-u)/(kn(i+k)-kn(i+1)))*B(i+1,k-1);
22              end
23          end
24      end
25  end
```

and the remaining basis functions can be calculated recursively from

$$B_{i,d}(u) = \frac{u - k^{(i)}}{k^{(i+d-1)} - k^{(i)}} B_{i,d-1}(u) + \frac{k^{(i+d)} - u}{k^{(i+d)} - k^{(i+1)}} B_{i+1,d-1}(u), \qquad (3.33)$$

noting that the term should be evaluated as zero where a zero denominator occurs. A function to calculate the B weightings is included in Listing 3.7. The inputs are the knot vector (kn), the degree of the NURBS and the scalar parameter $u \in [k^{(d)}, k^{(n+1)}]$, which denotes the position along the NURBS.

3.7   Implementation in Rhino

In Listings 3.2 and 3.3 we showed the construciton of Bézier curves using OpenNURBS/Rhino-Python's AddCurve. Now we will show two straightforward examples of using AddNurbsCurve (Curve being a base class of NurbsCurve) to create the range of splines covered in the preceding sections. When calling AddNurbsCurve(points,knots,degree,weights), a range of spline types can be created by varying the four attributes.

Listing 3.8 OpenNURBS/Rhino-Python to create the Bézier spline aerofoil in Figure 3.10.

```
1  import rhinoscriptsyntax as rs
2  # seven control points from trailing edge,
3  # to leading edge, to trailing edge
4  points=((1.0,   0.0,    0.0),
5          (0.5,   0.08,   0.0),
6          (0.0,  -0.05,   0.0),
7          (0.0,   0.025,  0.0),
8          (0.0,   0.1,    0.0),
9          (0.4,   0.2,    0.0),
10         (1.0,   0.0,    0.0))
11 degree=3
12 # repeated knots so as to pass through
13 # trailing and leading edge points
14 knots=(0, 0, 0, 1, 1, 1, 2, 2, 2)
15 # equal weighting of control points
16 weights=(1, 1, 1, 1, 1, 1, 1)
17 rs.AddNurbsCurve(points, knots, degree, weights)
```

We have shown how to create the aerofoil in Figure 3.10 using a Bézier spline by finding the joining point of two degree-3 Bézier curves. We can instead create this spline using a NURBS of degree 3, with repeated knots at the trailing and leading edge points. Referring to Figure 3.10, we have seven control points (note there are two at the trailing edge – the first and last, $\mathbf{a}^{(0)}$ and $\mathbf{a}^{(6)}$), and so there are $(n+1)+(d+1)=7+4=11$ knots. However, Rhino uses an $(n+1)+d-1$ knot vector (nine knots). The implementation is the same, but d, rather than $d+1$, knot repetitions are specified to force the NURBS to start and end at the first and last control points. With $d=3$ repetitions at the trailing edge points and $d=3$ repetitions required for the NURBS to interpolate the leading edge point, the knot vector is $0,0,0,1,1,1,2,2,2$. This produces a spline with two degree-3 segments (i.e. an upper and lower Bézier curve). The OpenNURBS/Rhino-Python script is shown in Listing 3.8. The `weights` are all set to one for this Bézier spline, but added shape control is achieved by varying this vector (a *rational* Bézier spline); for example, by increasing the value of the fourth (leading edge) control point weight.

The increased local control near the trailing edge in Figure 3.17 was obtained with an upper Bézier curve and a lower NURBS, and its OpenNURBS/Rhino-Python implementation is shown in Listing 3.9.

3.8 Curves for Optimization

This is a book about geometry for optimization and, as such, we will conclude this chapter with a few comments about how the curves of the preceding sections fit into the discussion in Sections 2.2 and 2.3. Here, we have covered a range of increasingly flexible generic curves, which has gone hand in hand with an increasingly complex parameterization. In Chapters 6 and 7 we consider special cases of lifting surfaces and their concise parameterization. Clearly, conciseness is in tension with flexibility of parameterization.

Listing 3.9   OpenNURBS/Rhino-Python to create the Bézier-curve/NURBS aerofoil in Figure 3.17.

```
1  import rhinoscriptsyntax as rs
2  # four upper control points from trailing edge,
3  # to leading edge
4  upperpoints=((0.0, 0.025, 0.0),
5             (0.0, 0.1, 0.0),
6             (0.4, 0.2, 0.0),
7             (1.0, 0.0, 0.0))
8  degree=3
9  # repeated knots so as to pass through
10 # trailing and leading edge points
11 knots=(0, 0, 0, 2, 2, 2)
12 # equal weighting of control points
13 weights=(1, 1, 1, 1)
14 rs.AddNurbsCurve(upperpoints, knots, degree, weights)
15
16 # five lower control points from trailing edge,
17 # to leading edge
18 lowerpoints=((1.0, 0.0, 0.0),
19               (0.8, -0.0, 0.0),
20               (0.5, 0.0, 0.0),
21               (0.0, -0.05, 0.0),
22               (0.0, 0.025, 0.0))
23 degree=2;
24 # repeated knots so as to pass through
25 # trailing and leading edge points
26 knots=(0, 0, 1, 2, 3, 3)
27 weights=(1, 1, 1, 1, 1)
28 obj=rs.AddNurbsCurve(lowerpoints, knots, degree, weights)
```

The key to a successful design process (in terms of obtaining a good design at minimal cost) is to incorporate as little flexibility as possible. However, this is seldom the mantra of the budding engineer with all the power of a 3D CAD engine at their fingertips, creating ever more complex NURBS representations of, what might be, far simpler geometry. Piegl and Tiller (1996) hoped that their book would put an end to the joke acronym 'nobody understands rational basis-splines'. In the meantime, the problem is perhaps as much that *everybody* uses NURBS even though nobody understands them.

Looking again at Figure 3.19, we see that, by rows (e) and (f), where all the flexibility of NURBS is on show, the geometry is not likely to be useful in aerodynamic design and the number of variables defining it would make for a difficult-to-solve optimization problem, were it a parametric geometry to be optimized. Rows (b) (a Bézier curve) and (c) (nonuniform only to force the spline to start and end at the first and last control points) look like better candidates for aircraft geometry. Naturally, the complexity of the parameterization depends on the problem at hand, but we would encourage starting at the beginning of this chapter and working forwards, rather than heading for NURBS with all the 'bells and whistles' straight away; *entia non sunt multiplicanda sine necessitate....*

4

Surfaces

4.1 Lofted, Translated and Coons Surfaces

The extension to surfaces of the various curves and splines covered in the previous sections is mercifully straightforward. In this section all the surfaces are based on Bézier curves, but could equally be based on the B-spline or NURBS definitions of the previous sections.

Lofted surfaces are perhaps the simplest (and also perhaps one of the most common), and are simply a weighted sum of two curves:

$$\mathcal{L}(u, t) = (1 - u)B^{(1)}(t) + uB^{(2)}(t), \tag{4.1}$$

where $B^{(\cdot)}(t)$ are curves traced out by varying t, and u is the parameter that moves us from curve $B^{(1)}(t)$ to $B^{(2)}(t)$ in a straight line. A lofted surface between two Bézier curves is shown in Figure 4.1. An OpenNURBS/Rhino-Python implementation using `AddLoft()` is shown in Listing 4.1.

Rather than loft directly from one curve to the next, we can translate one curve along another intersecting curve, giving us, for example, the freedom to create curved leading edges. Such a surface is expressed as the sum of the two intersecting curves, minus the intersection point:

$$\mathcal{T}(u, t) = B^{(1)}(u) + B^{(2)}(t) - \mathbf{a}^{(0,0)}, \tag{4.2}$$

where $\mathbf{a}^{(0,0)}$ is the point of intersection. Note that here u moves along the surface in the direction of the first curve and t in the direction of the second. Figure 4.2 shows such a surface created with two Bézier curves, which can be implemented with OpenNURBS/Rhino-Python using `AddSweep1()`.

The capabilities of lofted and translated surfaces are combined in the Coons surface, which is defined by four boundary curves. The surface is created in a similar way to a translational surface: as the sum of two lofts between opposite boundary curves, minus a weighted sum of the corner points. The weighted sum is in fact a surface in itself – a bilinear surface:

$$\mathcal{BL}^{(ab)}(u, t) = (1 - u, \ u) \begin{pmatrix} \mathbf{a}^{(0,0)} & \mathbf{a}^{(0,m)} \\ \mathbf{a}^{(n,0)} & \mathbf{a}^{(n,m)} \end{pmatrix} \begin{pmatrix} 1 - t \\ t \end{pmatrix}. \tag{4.3}$$

Aircraft Aerodynamic Design: Geometry and Optimization, First Edition. András Sóbester and Alexander I J Forrester.
© 2015 John Wiley & Sons, Ltd. Published 2015 by John Wiley & Sons, Ltd.

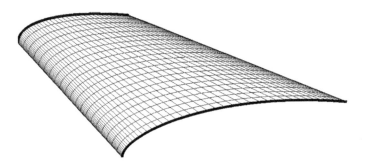

Figure 4.1 Lofted surface between two Bézier curves.

Listing 4.1 OpenNURBS/Rhino-Python to create the Bézier curve-based loft in Figure 4.1 (figure is created in MATLAB®).

```
1  import rhinoscriptsyntax as rs
2  # inner aerofoil control points
3  inner_points=((0.0, 0.0, 0.0),
4                (0.0, 0.0, 0.1),
5                (0.4, 0.0, 0.2),
6                (1.0, 0.0, 0.0))
7  # outer aerofoil control points
8  outer_points=((0.0, 1.0, 0.0),
9                (0.0, 1.0, 0.1),
10               (0.2, 1.0, 0.1),
11               (0.5, 1.0, 0));
12 # compute inner and outer degree
13 # three Bezier
14 degree=3
15 inner= rs.AddCurve(inner_points,degree)
16 outer=rs.AddCurve(outer_points,degree)
17 # add lofter surface
18 rs.AddLoftSrf(inner,outer)
```

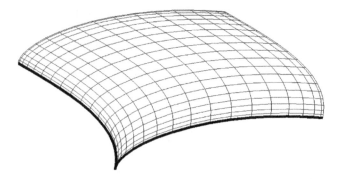

Figure 4.2 Surface created by translating one Bézier curve along a second, intersecting Bézier curve.

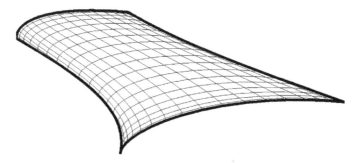

Figure 4.3   Coons surface created with four boundary Bézier curves.

Calculating $\mathcal{L}^{(a)}(u,t)$ and $\mathcal{L}^{(b)}(u,t)$ from Equation 4.1, the Coons surface is expressed as

$$C(u,t) = \mathcal{L}^{(a)}(u,t) + \mathcal{L}^{(b)}(u,t) - B\mathcal{L}^{(ab)}(u,t). \tag{4.4}$$

A Coons surface based on four boundary Bézier curves is shown in Figure 4.3. For this wing-type shape, note the ability to manipulate root and tip profiles and leading and trailing edge paths. An OpenNURBS/Rhino-Python implementation using `AddSweep2()` is shown in Listing 4.2.

The three forms of surface described in this section are all variations on a theme of using curves to define surfaces. This is a powerful method, but sometimes more local control of the surface may be desired, and this can be obtained by manipulating points that define the surface. We will now go on to consider surfaces created from grids of control points, which, although more awkward to define, yield enhanced local control.

4.2   Bézier Surfaces

We will begin with taking the Bézier curve, Equation 3.12 or 3.14, which is traced out as parameter u is varied, and creating a Bézier surface for varying u and t. This surface can be considered as a grid of curves, with each traced out by t or u for a fixed u or t, as shown in Figure 4.4. The surface is defined using an $n \times m$ grid of control points $\mathbf{a}^{(0,\ldots,n,0,\ldots,m)}$, and just as a Bézier curve starts and ends at the first and last control points, so does a Bézier surface have the grid corner control points at its corners. In Figure 4.4, the gridline for $u = 0, t \in 0, 1$ is shown in bold to highlight its dependence on the control points $\mathbf{a}^{(0,0)}$, $\mathbf{a}^{(1,0)}$ and $\mathbf{a}^{(2,0)}$; that is, it is a simple Bézier curve.

A point on the surface is expressed as the sum of the product of each control point and its weighting in both the u and t directions:

$$S(u,t) = \sum_{i=1}^{n} \sum_{j=1}^{m} \mathbf{a}^{(i,j)} b_i(u) f_j(t). \tag{4.5}$$

Listing 4.2 OpenNURBS/Rhino-Python to create the Bézier curve-based Coons surface in Figure 4.3 (figure is created in MATLAB®).

```
 1  import rhinoscriptsyntax as rs
 2  # aerofoil control points
 3  inner_points=((0.0, 0.0, 0.0),
 4                (0.0, 0.0, 0.1),
 5                (0.4, 0.0, 0.2),
 6                (1.0, 0.0, 0.0))
 7  outer_points=((0.0, 1.0, 0.0),
 8                (0.0, 1.0, 0.1),
 9                (0.2, 1.0, 0.1),
10                (0.5, 1.0, 0))
11  # leading and trailing edge control points
12  le_points=((0.0, 0.0, 0.0),
13             (0.2, 0.33, 0.0),
14             (0.1, 0.66, 0.0),
15             (0.0, 1.0, 0))
16  te_points=((1.0, 0.0, 0.0),
17             (0.8, 0.33, 0.0),
18             (0.6, 0.66, 0.0),
19             (0.5, 1.0, 0.0))
20  # compute boundary Bezier curves
21  degree=3
22  inner= rs.AddCurve(inner_points,3)
23  outer= rs.AddCurve(outer_points,3)
24  le=rs.AddCurve(le_points,3)
25  te=rs.AddCurve(te_points,3)
26  # add 'Coons' surface
27  rs.AddSweep2((le, te), (inner, outer))
```

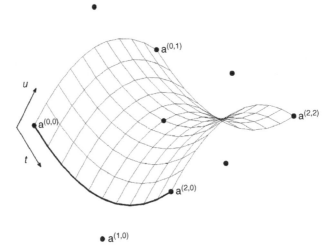

Figure 4.4 A simple Bézier surface based on a 3×3 grid of control points. Considered as a grid of Bézier curves, one such curve is shown in bold and is traced out by varying t, with u held constant.

The $b_i(u)$ and $f_j(t)$ are the usual Bernstein polynomials (Equation 3.13). As for a curve, the surface can also be calculated in matrix form as

$$S(u,t) = (t^n, t^{n-1}, \ldots, t, 1)\mathbf{N}^{(n)}\mathbf{A}\mathbf{M}^{(m)\mathrm{T}}(u^m, t^{m-1}, \ldots, u, 1)^{\mathrm{T}}, \qquad (4.6)$$

where the $\mathbf{N}^{(n)}$ and $\mathbf{M}^{(m)}$ Bernstein polynomial coefficient matrices are given in Equation 3.16 and

$$\mathbf{A} = \begin{pmatrix} \begin{pmatrix} a_1^{(1,1)} \\ a_2^{(1,1)} \\ a_3^{(1,1)} \end{pmatrix} & \begin{pmatrix} a_1^{(1,2)} \\ a_2^{(1,2)} \\ a_3^{(1,2)} \end{pmatrix} & \cdots & \begin{pmatrix} a_1^{(1,m)} \\ a_2^{(1,m)} \\ a_3^{(1,m)} \end{pmatrix} \\[2em] \begin{pmatrix} a_1^{(2,1)} \\ a_2^{(2,1)} \\ a_3^{(2,1)} \end{pmatrix} & \begin{pmatrix} a_1^{(2,2)} \\ a_2^{(2,2)} \\ a_3^{(2,2)} \end{pmatrix} & \cdots & \begin{pmatrix} a_1^{(2,m)} \\ a_2^{(2,m)} \\ a_3^{(2,m)} \end{pmatrix} \\[2em] \vdots & \vdots & \ddots & \vdots \\[1em] \begin{pmatrix} ca_1^{(n,1)} \\ a_2^{(n,1)} \\ a_3^{(n,1)} \end{pmatrix} & \begin{pmatrix} a_1^{(n,2)} \\ a_2^{(n,2)} \\ a_3^{(n,2)} \end{pmatrix} & \cdots & \begin{pmatrix} a_1^{(n,m)} \\ a_2^{(n,m)} \\ a_3^{(n,m)} \end{pmatrix} \end{pmatrix}.$$

The MATLAB® code in Listing 4.3 implements Equation 4.6. Note that a grid of 4×3 control points is defined, yielding the surface of degree 2 in the u-direction and degree 3 in the t-direction shown in Figure 4.5.

Above, we have defined what is known as a *Bézier patch*. A series of these can be joined to produce a complex surface – just like with Bézier splines – by calculating the corner points of the patches using Equation 3.20. Figure 4.6 shows a simple tapered wing with lower $\mathbf{L}$ and upper $\mathbf{U}$ control points, defined as

$$\mathbf{L}^{[(0,\ldots,n-1),(0,\ldots,m-2)]} = \begin{pmatrix} \begin{pmatrix} 1 \\ 0 \\ 0 \end{pmatrix} & \begin{pmatrix} 0.5 \\ 2.5 \\ 0 \end{pmatrix} \\[2em] \begin{pmatrix} 0.5 \\ 0 \\ 0.08 \end{pmatrix} & \begin{pmatrix} 0.25 \\ 2.5 \\ 0.04 \end{pmatrix} \\[2em] \begin{pmatrix} 0 \\ 0 \\ -0.05 \end{pmatrix} & \begin{pmatrix} 0 \\ 2.5 \\ -0.025 \end{pmatrix} \end{pmatrix}$$

and

$$U^{[(0,\ldots,n-1),(1,\ldots,m-1)]} = \begin{pmatrix} \begin{pmatrix} 0 \\ 0 \\ 0 \end{pmatrix} & \begin{pmatrix} 0 \\ 2.5 \\ 0.05 \end{pmatrix} \\ \begin{pmatrix} 0.4 \\ 0 \\ 0.2 \end{pmatrix} & \begin{pmatrix} 0.2 \\ 2.5 \\ 0.1 \end{pmatrix} \\ \begin{pmatrix} 1 \\ 0 \\ 0 \end{pmatrix} & \begin{pmatrix} 0.5 \\ 2.5 \\ 0 \end{pmatrix} \end{pmatrix},$$

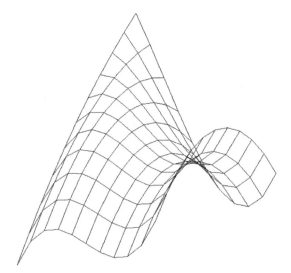

Figure 4.5   Bézier surface produced by the code in Listing 4.3.

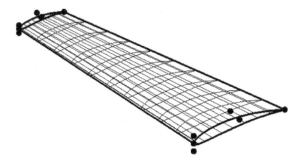

Figure 4.6   A simple wing created with upper and lower Bézier patches.

Listing 4.3 MATLAB® code for generating a Bézier surface; that is, implementing Equation 4.6.

```matlab
 1  % n x m grid of points
 2  P(:,:,1)=[0 0.5 1
 3           0 0.5 1
 4           0 0.5 1
 5           0 0.5 1];
 6  P(:,:,2)=[0 0 0
 7           0.5 0.5 0.5
 8           1 1 1
 9           1.5 1.5 1.5];
10  P(:,:,3)=[0 1 0
11           -1 0 -1
12           0 1 0
13           -1 0 1];
14  % define u and t in[0,1]
15  u=0:0.1:1;
16  t=0:0.1:1;
17  % degree of surface in each direction
18  n=3;
19  m=2;
20  % calculate Bernstein polynomial coefficent matrices
21  N=beziermatrix(n);
22  M=beziermatrix(m);
23  % for and 11 x 11 grid ...
24  for l=0:10
25      for k=0:10
26          % calculate x,y,z coordinates of surface points in turn
27          bezierPoints(k+1,l+1,1)=[t(k+1)^3 t(k+1)^2 t(k+1),1]*...
28              N*P(:,:,1)*M'*[u(l+1)^2 u(l+1), 1]';
29          bezierPoints(k+1,l+1,2)=[t(k+1)^3 t(k+1)^2 t(k+1),1]*...
30              N*P(:,:,2)*M'*[u(l+1)^2 u(l+1), 1]';
31          bezierPoints(k+1,l+1,3)=[t(k+1)^3 t(k+1)^2 t(k+1),1]*...
32              N*P(:,:,3)*M'*[u(l+1)^2 u(l+1), 1]';
33      end
34  end
35  % plot mesh
36  m=mesh(bezierPoints(:,:,1),bezierPoints(:,:,2),bezierPoints(:,:,3));
```

with the start/end – that is, the leading edge join calculated using Equation 3.20 – as

$$
\mathbf{L}^{[(0,\dots,n-1),(m-1)]} = \left(\begin{pmatrix} 0 \\ 0 \\ 0.025 \end{pmatrix} \begin{pmatrix} 0 \\ 2.5 \\ 0.0125 \end{pmatrix} \right)
$$

and

$$
\mathbf{U}^{[(0,\dots,n-1),(0)]} = \left(\begin{pmatrix} 0 \\ 0 \\ 0.025 \end{pmatrix} \begin{pmatrix} 0 \\ 2.5 \\ 0.0125 \end{pmatrix} \right).
$$

This linearly tapered wing could be created much more easily as two lofted surfaces (Equation 4.1), but serves here for illustrative purposes. As with curves, local control and ease of implementation can be found in the use of a B-spline approach.

4.3   B-Spline and Nonuniform Rational Basis Spline Surfaces

A B-spline surface is generated as a grid of patches, calculated as the sum of the products of B-spline curves defined over a grid of control points:

$$
\begin{aligned}
S_{i,j}(u,t) &= (t^n, t^{n-1}, \dots, t, 1)\mathbf{N}^{(n)} \\
&\times \mathbf{A}^{[(i-1,\dots,i+n-1),(j-1,\dots,j+m-1)]}\mathbf{M}^{(m)\mathrm{T}}(u^m, t^{m-1}, \dots, u, 1)^{\mathrm{T}};
\end{aligned}
\tag{4.7}
$$

that is, the relationship between B-spline and Bézier curves is much the same as that between their surface counterparts. Figure 4.7 shows a quadratic B-spline surface defined by a 7×7 grid of control points. A quadratic B-spline curve defined by seven control points would have five segments, and so here we have a grid of 5×5 patches, whose edges are shown in bold.

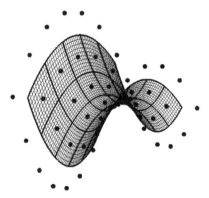

Figure 4.7   B-spline surface based on a 7×7 grid of control points.

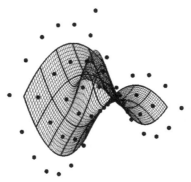

Figure 4.8 B-spline surface based on a 7×7 grid of control points, with local control of the central region by moving the centre control point $\mathbf{a}^{(3,3)}$ with respect to Figure 4.7.

In Figure 4.8 we see the local control achieved by moving the centre control point; this distorts a 3×3 grid of patches around the point, but leaves the remainder of the surface unchanged.

A rational B-spline surface has a weighting associated with each control point, with a point on the surface now calculated as

$$S_{i,j}(u,t) = \frac{(t^n, t^{n-1}, \ldots, t, 1)\mathbf{N}^{(n)}[\mathbf{A}^{[(i-1,\ldots,i+n-1),(j-1,\ldots,j+m-1)]} \bullet \mathbf{Z}^{[(i-1,\ldots,i+n-1),(j-1,\ldots,j+m-1)]}]\mathbf{M}^{(m)\mathrm{T}}(u^m, t^{m-1}, \ldots, u, 1)^{\mathrm{T}}}{(t^n, t^{n-1}, \ldots, t, 1)\mathbf{N}^{(n)}\mathbf{Z}^{[(i-1,\ldots,i+n-1),(j,\ldots,j+m-1)]} \times \mathbf{M}^{(m)\mathrm{T}}(u^m, t^{m-1}, \ldots, u, 1)^{\mathrm{T}}}, \qquad (4.8)$$

where $\mathbf{Z}$ is an $n \times m$ matrix of control point weights and here $\bullet$ denotes an element-by-element product. Figure 4.9 shows the effect of increasing the weighting of the centre control point, distorting the local surface towards this control point.

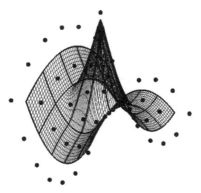

Figure 4.9 Rational B-spline surface based on a 7×7 grid of control points, with central region distorted by increasing the weight of point $\mathbf{a}^{(3,3)}$ with respect to Figure 4.7.

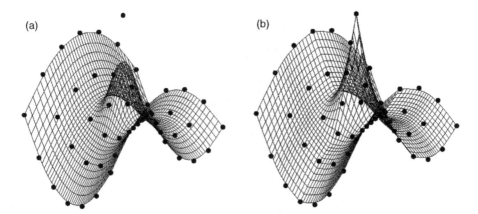

Figure 4.10 NURBS surfaces where repeated knot values force the surface to interpolate: (a) a set of the edge control points and (b) all of the edge control points and the centre control point, where there is a spike.

The advantages of NURBS were discussed in Section 3.6, and the flexibility of employing a nonuniform knot vector was highlighted in Figure 3.19. NURBS surfaces offer the same benefits. A NURBS patch is calculated similarly to a B-spline:

$$\mathcal{N}_{i,j}(u, t) = \frac{[\mathbf{z} \times \mathbf{N}^{(n)}(t)]\mathbf{A}[\mathbf{z} \times \mathbf{M}^{(m)}(u)]^{\mathrm{T}}}{\mathbf{z}\mathbf{N}^{(n)}\mathbf{M}^{(m)}\mathbf{z}^{\mathrm{T}}}, \tag{4.9}$$

but with the recursive definition of the weights $B_{i,d}(t)$ detailed in Section 3.6 to create the $\mathbf{N}^{(n)}(t)$ and $\mathbf{M}^{(m)}$ weight matrices. Figure 4.10 (a) shows the use of a nonuniform knot vector $(0, 0, 0, 1, 2, 3, 4, 5, 5, 5)$, where repeated knots force the degree-2 surface edges to the edges of the 7×7 control point grid. In Figure 4.10 (b) the knot vector $(0, 0, 0, 1, 2, 2, 3, 4, 4, 4,)$ also forces the NURBS surface to interpolate the centre control point. In both cases the knot vector is the same in both the u and t directions.

4.4   Free-Form Deformation

Free-form deformation (FFD), which was first introduced by Sederberg and Parry (1986), is a process by which wholesale shape changes can be made to geometry by manipulating the locations of points that are related to the geometry. It is a powerful, yet surprisingly simple method based on concepts we have already covered when constructing Bézier curves (Section 3.1). We will first look at FFD of 2D geometry, which we did not cover in Chapter 3 as FFD tends to be applied to 3D geometry, and here we consider 2D purely as an introduction to 3D FFD. A number of enhancements to FFD have been suggested since Sederberg and Parry's formulation, which we cover here, and the reader may wish to refer to Coquillart (1990) for details of 'extended FFD' and Hsu et al. (1992) for 'directly manipulated FFD'.

   FFD manipulates geometry by attaching it to Bézier curves, which are defined by a grid of control points defined over $u \in [0, 1], t \in [0, 1]$. Such an $n \times m$ grid of control points $\mathbf{a}^{(i,j)}$ is shown in Figure 4.11. The control points' influence on the space within the grid is governed by

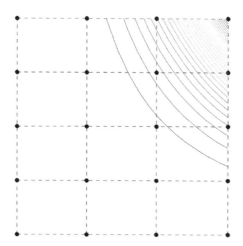

Figure 4.11   FFD control point grid and contour map of the $b_n(u)f_m(t)$ Bernstein polynomial.

the Bernstein polynomials (see Figures 3.5 and 3.6). As with Bézier curves and surfaces, the end/edge control points are the sole contributors at their respective ends/edges. The $n \times m$th Bernstein polynomial $b_n(u)f_m(t)$ is shown as a contour map over the grid in Figure 4.11. Its maximum influence is at $u, t = 1$ and it has zero influence at $u, t = 0$. Shifting the $\mathbf{a}^{(n,m)}$ control point will affect points within the grid according to this contour map. Figure 4.12 shows the affect of moving the control point (to $(0.75, 1.5)$ in Cartesian coordinates, but still at $u, t = 1$). Moving control points within the grid will have less dramatic effects, as shown in Figure 4.13.

This FFD, based on a grid of control points, is expressed mathematically as

$$\mathbf{X}(u, t) = \sum_{i=0}^{n} \sum_{j=0}^{m} \mathbf{a}^{(i,j)} f_i(u) g_j(t), \tag{4.10}$$

where $\mathbf{X}(u, t)$ is the Cartesian coordinates of the new, deformed location of a point at (u, t). This mix of Cartesian coordinates and grid parameters is a little confusing. Consider the point at $x, y = 0.95, 0.85$ denoted by a cross in Figure 4.12. Its grid parameter is $u, t = 0.95, 0.85$. After the FFD, the point remains at $u, t = 0.95, 0.85$, but its Cartesian coordinates are shifted to $x, y = 0.8084, 1.1813$.

The FFD applies to all $u, t \in [0, 1]$ values. All we need to do is define a shape in grid parameter coordinates and apply Equation 4.10; it is a beautifully simple method. Figure 4.14a shows an FFD of a circle based on a 3×3 grid of control points (i.e. the FFD is controlled by $n, m = 3$ Bernstein polynomials). More complex geometry, text or images could equally be deformed. Only the geometry within the grid of control points is deformed (as shown in Figure 4.14b), which means that grids can be laid over a geometry of particular interest for design studies. First-order continuity with respect to adjacent, undeformed regions can be maintained with degree 2 or greater Bernstein polynomials; that is, $n, m >= 3$ in Equation 4.10.

Equation 4.10 is actually the same as Equation 4.5, but the grid of control points $\mathbf{a}^{(i,j)}$ has just two x, y coordinates. In the same way as increased flexibility can be added to Bézier curves

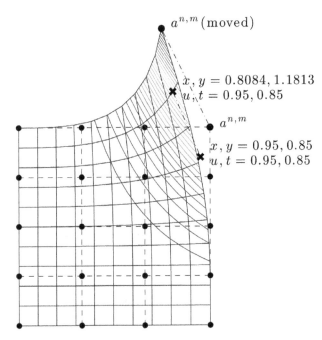

Figure 4.12 FFD control point grid with the $\mathbf{a}^{(n,m)}$ control point moved. The solid lined grid shows how points are shifted according to the influence of the control points. The $b_n(u)f_m(t)$ Bernstein polynomial is shown as a contour map.

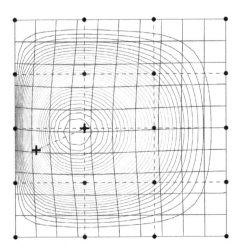

Figure 4.13 FFD control point grid with the $\mathbf{a}^{(2,3)}$ control point moved. The solid lined grid show how points are shifted according to the influence of the control points. The $b_2(u)f_3(t)$ Bernstein polynomial is shown as a contour map.

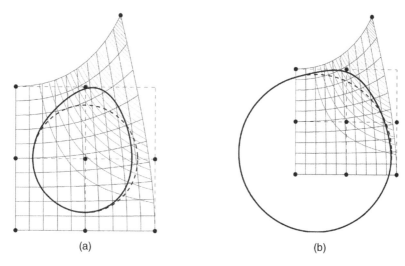

(a)   (b)

Figure 4.14   FFD of a circle by shifting $\mathbf{a}^{(n,m)}$ of a 3×3 grid encompassing (a) the whole circle and (b) laid over part of the circle.

and surfaces by using weighted control points, and then on to B-Splines and NURBS, similar additional local control may be added to FFD. However, we will consider only the elegant Bernstein polynomial form of FFD.

Extending FFD to three dimensions is straightforward, requiring a 3D grid of control points and a third set of Bernstein polynomials:

$$\mathbf{X}(u,t,s) = \sum_{i=0}^{m}\sum_{j=0}^{n}\sum_{k=0}^{p}\mathbf{a}^{(i,j,j)}f_i(u)g_j(t)h_k(s). \qquad (4.11)$$

Listing 4.4 is a MATLAB® implementation of Equation 4.11 called ffdsurf(s,t,u,a). The inputs s, t and u are matrices of the grid parameter values of a surface to be deformed. This matrix format is in line with other MATLAB® functions, such as surf(). The fourth input is the grid of control points defining the FFD. This is in the format produced by MATLAB®'s meshgrid(); that is, an $m \times n \times p \times 3$ matrix. Listing 4.5 is an example of using ffdsurf() to produce the deformed sphere shown in Figure 4.15b. If s, t and u are vectors, ffd() can be called and outputs a matrix with three x, y, z coordinate columns.

As a final example of the power of this simple method, Figure 4.16 shows a generic wing that has been deformed, using ffdsurf(), by the rotation and translation of the control point grid face at the wing-tip. If we had generated a computational fluid dynamics (CFD) mesh around our wing, this, too, could have been deformed, eliminating the need to remesh during an optimization process or, for example, when performing aero-elastic simulations, when the geometry could be deformed according to the effect of aerodynamic loads. Any entity described by s, t, u coordinates can, in effect, be made parametric by casting a grid over it and applying Equation 4.11.

Listing 4.4 MATLAB® function for computing the FFD of a grid of surface points using Equation 4.11.

```matlab
1  function X=ffdsurf(s,t,u,a)
2  % calculate degree of Bernstein polynomials
3  m=size(a,1)-1;
4  n=size(a,2)-1;
5  p=size(a,3)-1;
6  % pre-allocate memory
7  [ny,nx]=size(s);
8  X=zeros(ny,nx,3);
9  term=zeros(m+1,n+1,p+1,3);
10 f=zeros(m+1);
11 g=zeros(n+1);
12 h=zeros(p+1);
13 for ll=1:nx
14     for l=1:ny
15         for k=0:p
16             for j=0:n
17                 for i=0:m
18                     % calculate Bernstein polynomials
19                     f(i+1)=nchoosek(m,i)*t(l,ll)^i*(1-t(l,ll))^(m-i);
20                     g(j+1)=nchoosek(n,j)*s(l,ll)^j*(1-s(l,ll))^(n-j);
21                     h(k+1)=nchoosek(p,k)*u(l,ll)^k*(1-u(l,ll))^(p-k);
22                     % multiply by control points
23                     term(i+1,j+1,k+1,:)=f(i+1)*g(j+1)*h(k+1)...
24                     *a(i+1,j+1,k+1,:);
25                 end
26             end
27         end
28         % sum terms
29         X(l,ll,:)=squeeze(sum(sum(sum(term))));
30     end
31 end
```

Listing 4.5 MATLAB® function to create the FFD sphere in Figure 4.15 by calling `ffdsurf()` in Listing 4.4.

```matlab
1  % 2x2x2 grid of control points
2  [a(:,:,:,1),a(:,:,:,2),a(:,:,:,3)]=meshgrid([0 1],[0 1],[0 1]);
3  % move corner control point
4  a(2,2,2,:)=[1.5 1.5 1.5];
5  % create sphere
6  [s,t,u]=sphere(20);
7  % shift to within s,t,u grid
8  s=(s+1)./2;
9  t=(t+1)./2;
10 u=(u+1)./2;
11 % call ffdsurf
12 X=ffdsurf(s,t,u,a);
13 % display deformed sphere
14 m=mesh(X(:,:,1),X(:,:,2),X(:,:,3));
```

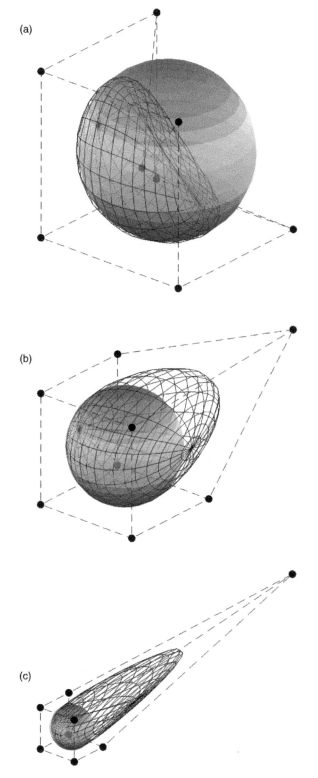

(a)

(b)

(c)

Figure 4.15 FFD of a sphere by shifting $\mathbf{a}^{(m,n,p)}$ of a $2 \times 2 \times 2$ grid.

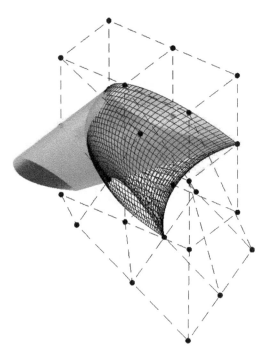

Figure 4.16 FFD of a generic wing shape, with the control points on the face of the wing-tip end of the grid rotated and translated.

4.5 Implementation in Rhino

4.5.1 *Nonuniform Rational Basis Splines-Based Surfaces*

The OpenNURBS/Rhino-Python script `AddNurbsSurface()` can, in a similar way to its curve counterpart `AddNurbsCurve()`, be used to create a range of surfaces, from Bézier to NURBS, by controlling the degree and knot vectors. Listing 4.6 is an OpenNURBS/Rhino-Python implementation, equivalent to Listing 4.3, and produces the Bézier surface in Figure 4.17, which is equivalent to Figure 4.5.

Figure 4.6 is constructed from two Bézier surface patches. Listing 4.7 uses `AddNurbsSurface()` with the aerofoil cross-section curves as degree-3 NURBS in the t direction, with repeated control points to force the curve to start and end at the trailing edge and pass through the leading edge. In the u direction, a straight wing is produced with degree 1. The Rhino geometry is shown in Figure 4.18.

The NURBS surfaces in Figure 4.10 show the effect of repeating knot values. Listing 4.8 is an OpenNURBS/Rhino-Python implementation, and the Rhino output is shown in Figure 4.19.

4.5.2 *Free-Form Deformation*

The simplicity of FFD means that geometry can easily manipulated outside of a CAD tool, facilitating integration with simulation and optimization tools; for example, the mesh can be

Listing 4.6   OpenNURBS/Rhino-Python script to create the Bézier surface in Figure 4.4.

```
 1  import rhinoscriptsyntax as rs
 2  import Rhino
 3  # coordinates of control points
 4  point_coords=((0, 0, 0),
 5                (0.5, 0, 1),
 6                (1, 0, 0),
 7                (0, 0.5, -1),
 8                (0.5, 0.5, 0),
 9                (1, 0.5, -1),
10                (0, 1, 0),
11                (0.5, 1, 1),
12                (1, 1, 0),
13                (0, 1.5, -1),
14                (0.5, 1.5, 0),
15                (1, 1.5, 0.5))
16  # initialise Rhino 3D points list
17  points = Rhino.Collections.Point3dList(12)
18  # add points to list
19  for i in range(0,12):
20      points.Add(rs.coerce3dpoint(point_coords[i]))
21  # no. of points in u,t directions
22  point_count=(4, 3)
23  # degree of surface in u,t directions set at
24  # n,m-1 to give Bezier surface
25  degree=(3, 2)
26  # knot vectors with repeated knots to
27  # give Bezier surface
28  knots_u=(0, 0, 0, 1, 1, 1)
29  knots_t=(0,0,1, 1)
30  # all control point weights set at 1 for
31  # non-rational Bezier surface
32  weights=(1, 1, 1, 1, 1, 1, 1, 1, 1, 1, 1, 1)
33  # AddNurbsSurface to model
34  obj = rs.AddNurbsSurface(point_count, points, knots_u, knots_t,\
35    degree, weights)
```

deformed at the same time. However, there are clear advantages to manipulating the geometry in the CAD tool, as CAD lies at the centre of many design processes.

Many CAD tools have FFD functionality through the graphical user interface. Indeed, the intuitive user interaction with FFD is one of its strengths (e.g. in the animation industry). However, for the purpose of engineering design, automated scripting is essential. Listing 4.9 is an example of an OpenNURBS/Rhino-Python script that could be used to perform the FFD in Figure 4.15. The Rhino output is shown in Figure 4.20.[1]

---

[1] The use of a macro to implement the Rhino Cage Edit function is rather clumsy. Functionality is constantly being added to Python for Rhino, but at the time of publication a simple RhinoScript implementation for Cage Edit was not available.

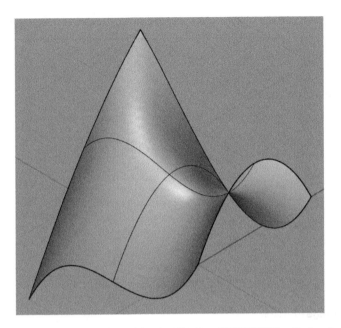

Figure 4.17   A reproduction of Figure 4.5 using the OpenNURBS/Rhino-Python in Listing 4.6.

Figure 4.21 is an example of a more impressive, yet equally simple, implementation of FFD. FFD 'cages' have been created around the port wing-tip and starboard wing of 'SUHPA' (whose wing design is the subject of Chapter 12). For the starboard wing-tip, two control points (degree 1) are used in the x and z (upwards and aft) directions, and four control points (degree 2) in the y (span-wise) direction. The end face of the FFD cage has then been rotated 90° about the next inboard set of lower cage control points. The result is a blended winglet, which can easily be modified further, in terms of length, angle and sweep, and so on, by moving the FFD cage control points.[2] The port wing has been twisted by rotating the end face of the FFD cage about its lower forward control point to produce an exaggerated tip washout. Further reading in the area of CAD/FFD/CFD includes Nurdin *et al.* (1183), who present an example of an FFD-based wing/fuselage fillet optimization using CATIA® and Fluent®.

4.6   Surfaces for Optimization

As we did for curves, we will end this introduction to some of the fundamentals of generating and manipulating parametric surfaces with a note on their suitability for aerodynamic design optimization.[3] From lofts through to NURBS surfaces, we have moved from shapes defined purely by their 2D cross-section, to fully 3D definitions. When choosing the type of

---

[2] We discuss another way of creating blended winglets in Section 9.3.2.

[3] There is a wider range of surface types available, and Salomon (2006) is an excellent resource on further curves and surfaces.

Listing 4.7 OpenNURBS/Rhino-Python to create the NURBS wing in Figure 4.18.

```
1  import rhinoscriptsyntax as rs
2  import Rhino
3  # coordinates of control points
4  point_coords=((1.0, 0.0, 0.0),
5              (0.5, 0.0, 0.08),
6              (0.0, 0.0, -0.05),
7              (0.0, 0.0, 0.025),
8              (0.0, 0.0, 0.1),
9              (0.4, 0.0, 0.2),
10             (1.0, 0.0, 0.0),
11             (0.5, 2.0, 0.0),
12             (0.25, 2.0, 0.08),
13             (0.0, 2.0, -0.05),
14             (0.0, 2.0, 0.025),
15             (0.0, 2.0, 0.1),
16             (0.2, 2.0, 0.2),
17             (0.5, 2.0, 0.0))
18 # initialise Rhino 3D points list
19 points = Rhino.Collections.Point3dList(14)
20 # add points to list
21 for i in range(0,14):
22     points.Add(rs.coerce3dpoint(point_coords[i]))
23 # display points
24 rs.AddPoints(points)
25 # no. of points in u,t directions
26 point_count=(2, 7)
27 # degree of surface in u,t directions
28 degree=(1, 3)
29 # knot vectors with repeated knots to start
30 # and end at TE and pass through LE
31 knots_u=(0, 1)
32 knots_t=(0, 0, 0, 1, 1, 1, 2, 2, 2)
33 # AddNurbsSurface to model
34 obj = rs.AddNurbsSurface(point_count, points, knots_u, knots_t,\
35  degree)
```

parameterization, we therefore need to consider whether the flow is largely 2D or whether it is dominated by 3D effects.

Consider, for example, the sailplane in Figure 1.1. This conventional wing–body–tail configuration is such that the flow over the wings and tail is largely 2D; that is, the major flow velocity is in the chord-wise direction, with little flow along the lifting surfaces. The majority of the fuselage is dominated by the fore–aft flow. The parts of the aircraft where the flow vectors become highly 3D, and so should the geometry definition, are the nose, wing–fuselage junction and tail–fuselage junction. It is natural, therefore, to employ basic lofted surfaces for

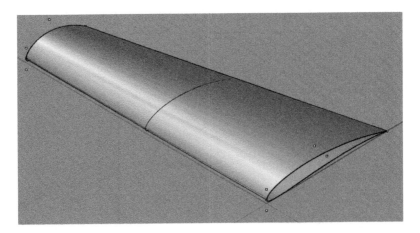

Figure 4.18 A reproduction of Figure 4.6 using the OpenNURBS/Rhino-Python in Listing 4.7.

Listing 4.8 OpenNURBS/Rhino-Python script to create the NURBS surfaces in Figure 4.19 (with lines of the control point list omitted).

```
 1  import rhinoscriptsyntax as rs
 2  import Rhino
 3  # coordinates of control points
 4  point_coords=((-0.25, -0.25, 0.00),
 5                 (0.00, -0.25, 0.31),
 6                 ...
 7                 (1.00, 1.25, 0.31),
 8                 (1.25, 1.25, 0.00))
 9  # initialise Rhino 3D points list
10  points = Rhino.Collections.Point3dList(49)
11  # add points to list
12  for i in range(0,49):
13      points.Add(rs.coerce3dpoint(point_coords[i]))
14  # no. of points in u,t directions
15  point_count=(7, 7)
16  # degree of surface in u,t directions set at
17  degree=(2, 2)
18  # knot vectors
19  # EITHER to interpolate edge points
20  # (left Rhino screenshot):
21  knots_u=(0, 0,  1, 2, 3, 4, 5, 5)
22  knots_t=(0, 0,  1, 2, 3, 4, 5, 5)
23  # OR to interpolate edge and centre points
24  # (right Rhino screenshot):
25  knots_u=(0, 0, 1, 2, 2, 3, 4, 4)
26  knots_t=(0, 0, 1, 2, 2, 3, 4, 4)
27  # AddNurbsSurface to model
28  obj = rs.AddNurbsSurface(point_count, points, knots_u, knots_t,\
29   degree)
```

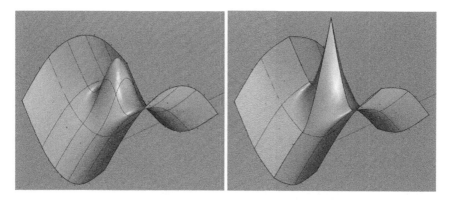

Figure 4.19 A reproduction of Figure 4.10 using the OpenNURBS/Rhino-Python in Listing 4.8.

Listing 4.9 OpenNURBS/Rhino-Python script to create the FFD sphere in Figure 4.20.

```
1  import rhinoscriptsyntax as rs
2  # draw sphere
3  obj=rs.AddSphere((0.5,0.5,0.5), 0.325)
4  # select sphere
5  rs.SelectObject(obj)
6  # macro to create FFD 'cage'
7  rs.Command ("_CageEdit BoundingBox World X 2 Y 2 Z 2 \
8  _Enter Global _Enter")
9  # select FFD cage
10 objcage=rs.ObjectsByType(131072,True)
11 # change location of cage point
12 point = rs.ObjectGripLocation(objcage, 7, (1.5, 1.5, 1.5))
```

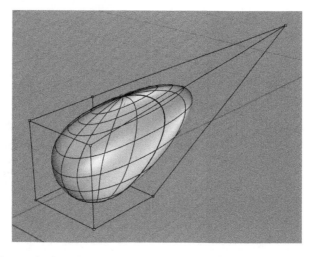

Figure 4.20 A reproduction of Figure 4.15 using the OpenNURBS/Rhino-Python in Listing 4.9.

Figure 4.21 FFD of the SUHPA geometry to create a blended winglet (starboard) and (exaggerated) tip washout (port) using Rhino's Cage Edit function.

the wings, tail and fuselage, with greater flexibility (e.g. from NURBS surfaces) at the nose and junctions. Section 2.5 further discusses the philosophy of conventional aircraft geometry modelling.

Of course, the sailplane in question is not an *optimal* aerodynamic shape, and the discretization of its geometry into these simple parts means that a design process would be limited to a conventional shape. More aerodynamically efficient, highly blended wing/body concepts, with stronger 3D flow effects, will benefit from more complex 3D surface definitions. As more control points are added and the flexibility of the surface increased, as we move from lofted, translated and Coons through to B-spline and NURBS surfaces defined by grids of control points, the complexity of manipulating surfaces increases, and so the usefulness to the designer diminishes; indeed, Figures 4.7–4.10 are examples of what *can* be done and not necessarily what *should* be done. As with curves, it is a case of finding the balance between limiting the shape too much and having so much flexibility that sensible designs will be lost in a vast design space.

5

Aerofoil Engineering: Fundamentals

The concept of an aerofoil ('airfoil' in the USA) has always been pivotal to humankind's attempts to engineer flying machines, and writings associated with it probably account for a greater percentage of the aeronautical engineering literature than any other single notion. 'The aerofoil section is the quintessence of a wing or lifting surface and, as such, occupies a central position in any design discipline relating to fluid mechanics, from animal flight through marine propellers to aircraft', notes Lissaman (1983).

The first and, possibly, last comprehensive monograph on the subject is a 700-page tome by Abbott and von Doenhoff (1959) (Dover's 1959 reprint is the best-known edition today, but the original was published in 1949). Their 'Theory of Wing Sections' is still a valuable reference today, in spite of the fact that it pre-dates by a long way the modern idea of aerofoil design based on numerical performance analysis.

Here, we shall make no attempt at an updated version of their book, as a similarly comprehensive work today would run to many thousands of pages. We concentrate instead on an aspect of aerofoil design that is most germane to the subject of this book: geometrical descriptions of aerofoils that enable parametric optimization studies. This chapter sets out some of the fundamental aerofoil-related ideas that any aerodynamics engineer should be familiar with, followed by a discussion of two categories of geometrical formulations of interest in analysis-driven aerodynamic design: aerofoil families (Chapter 6) and generic parametric formulations suitable for aerofoil description (Chapter 7).

5.1 Definitions, Conventions, Taxonomy, Description

An aerofoil section is the intersection of a lifting surface or a streamlined fairing and a plane. The term in itself gives limited indication of the positioning of the cutting plane. Sometimes this is taken to be parallel with the plane of symmetry of the pair of wings, tailplanes, canards, and so on (this is generally the same as the plane of symmetry of the aircraft). In the case of swept lifting surfaces and/or those with significant dihedral (we will define these concepts in

Aircraft Aerodynamic Design: Geometry and Optimization, First Edition. András Sóbester and Alexander I J Forrester.
© 2015 John Wiley & Sons, Ltd. Published 2015 by John Wiley & Sons, Ltd.

detail in Chapters 8 and 9 respectively; for now, we simply note that we mean that the leading edge is no longer a straight line perpendicular to the symmetry plane), the cutting plane may be orthogonal to the leading edge instead.

From an aerodynamic point of view it is generally considered good practice to choose the 'yaw' angle of the section plane to align with the streamlines of the flow (when viewed from above), though for reasons of geometrical expediency some engineers may prefer some other orientation.

Once the plane of the section is clarified, the next step is the selection of a Cartesian coordinate system, which will later enable us to position the aerofoil within the 3D geometry of the lifting surface. Once again, there are a number of possibilities, first in terms of choosing the origin.

The origin, or *leading edge point*, should, once again, ideally be chosen with an eye on the flow field around the section. This, however, is hindered somewhat by the fact that most such candidate locations (e.g. the surface point nearest to the stagnation point or the point on the surface that will 'see' the free stream first) vary with the *angle of attack* (the angle of incidence of the flow with respect to whatever datum direction we will go on to choose on the aerofoil). A 'special' angle of attack could be considered that yields zero lift on the aerofoil.[1] This would be a sensible convention to adhere to, but adherence to this principle can by no means be assumed when considering the coordinate sets describing published aerofoils.

The x-axis of the coordinate system defines the *chord* of the aerofoil, the imaginary line connecting the leading edge point with the trailing edge, though this leaves some room for ambiguity too. Sometimes, especially to simplify numerical simulations, the trailing edge is assumed to be sharp, in which case it is defined by a single point. However, in the case of a more realistic *finite-thickness trailing edge*, which point defines the aft end of the chord: the upper edge, the lower edge or perhaps one halfway between them? Once again, different practitioners opt for different definitions; fortunately, though, the impact of choosing one or the other is not great.

Another key feature of an aerofoil is its *camber*. This is a measure of the overall curvature of the aerofoil, and it is usually strongly linked to the lift coefficient: increasing the camber of an aerofoil means increasing the amount of lift it can generate at a particular angle of attack (typically at the expense of increasing the drag too).

The *camber curve* – sometimes also referred to as the *camber line* – is an imaginary curve connecting the endpoints of the chord. Its defining feature is that it is the geometrical locus of the points that are the same distance away from the upper and the lower surface of the aerofoil. In the spirit of the above it will come as no surprise that this is slightly open to interpretation too, as the distance could be defined along the z-axis (i.e. perpendicular to the chord) or perpendicular to the camber curve itself. See Figure 6.1 for an illustration of the latter definition.

5.2 A 'Non-Taxonomy' of Aerofoils

There is no unique, universally accepted classification for aerofoils, and we will not attempt one here either. Instead, we will consider a number of commonly used categories of lifting

---

[1] Of course, 2D lift is, in itself, a somewhat artificial concept.

surface sections based on shared traits like flow regime and the type of application they are designed for. Fittingly for a non-taxonomy, some aerofoils may fit into several of the categories below, others into none. The sections below may simply be viewed as starting points for a more in-depth study of the corresponding classes.

5.2.1   Low-Speed Aerofoils

Sometimes also termed *low Reynolds number aerofoils*, their design is generally driven by the desire to maximize the lift coefficient. This is either achieved through a pronounced camber or through a leading edge designed to allow flight at relatively high angles of attack.

While, as we noted, there is no hard and fast classification in terms of speed and/or Reynolds number, this category typically includes sections specifically designed for operation at Reynolds numbers below 500 000. Once largely the preserve of model aeroplane enthusiasts (who have made very significant contributions to their development), this category of profiles is experiencing renewed interest as a result of their use in light unmanned air systems.

The reader interested in more on their design and applications may wish to start with Lissaman (1983) and Selig (2003) as a good pair of primers on the subject.

5.2.2   Subsonic Aerofoils

These are perhaps the most numerous, as are their applications: those that involve free stream velocities up to about Mach 0.8. A common geometrical feature of these aerofoils is a *large leading edge radius*. This typically ensures robust behaviour; that is, the performance of the aerofoil does not exhibit large variations with small changes in angle of attack. At subsonic speeds this can be achieved without major drag sacrifice.

5.2.3   Transonic Aerofoils

These operate in subsonic, but *supercritical* free streams. The latter means that the local Mach number somewhere in the flow field exceeds 1.0. The *critical free-stream Mach number* at which this first occurs is usually around $0.7 \ldots 0.8$. This is typically followed closely by the *drag divergence free-stream Mach number* M_∞^{dd}; that is, the Mach number at which the formation of shock waves suddenly[2] increases the *wave drag*.

The maximization of the drag divergence Mach number is therefore a typical goal of transonic aerofoil optimization. Historically, this was typically achieved through an inverse design procedure, whereby a target pressure distribution with a flattened pressure peak and a gentle transition toward the aft end of the aerofoil was drawn up and the geometry was tailored to reproduce this target. An optimization process steered by a flow simulation trusted to yield a sufficiently accurate drag estimate could take the simpler route of minimizing the drag at a given Mach number or maximizing the drag rise Mach number (both, generally, computed by iterating the angle of attack until a target lift coefficient is met).

---

[2] A typical condition used to determine the location of this drag rise point is $\Delta C_{\mathrm{D}}/\Delta M)|_{M_\infty^{dd}} = 0.1$.

Figure 5.1   The Lockheed F-104 Starfighter: remarkable performance at the cost of unforgiving handling (photographs by A. Sóbester).

5.2.4   Supersonic Aerofoils

These operate at cruise conditions where the free-stream Mach number exceeds 1.0 (but it is less than about 5). Reducing wave drag in these conditions is rather challenging, and it can most readily achieved by reducing the thickness to chord ratio of the aerofoil, as well as by sharpening the leading edge.

A typical example of this design philosophy is the so-called *biconvex aerofoil*, featured, for instance, on the Lockheed F-104 Starfighter – see Figure 5.1. This aerofoil is a mere 3.36% chord thick! The F-104 is representative of the early supersonic jets designed with an overwhelming emphasis on high-speed performance; as a result, their low-speed handling often suffered significantly.

Not all supersonic aircraft feature thin wings with sharp leading edges. The presence of a relatively thin section with a somewhat rounded leading edge on a supersonic aircraft usually betrays a compromise between low wave drag at high Mach numbers and more robust performance (as well as better handling) at subsonic speeds. This sort of trade-off is typical of fast military jets, which are capable of a supersonic dash, but fly most of their missions at subsonic speeds.

5.2.5   Natural Laminar Flow Aerofoils

The shearing stresses in a boundary layer and, consequently, the skin friction drag are a function of the rate at which the velocity decays as we move closer to the surface. The steeper this gradient is, the greater the friction drag will be. In a laminar boundary layer, where there are no large-scale momentum exchanges between adjacent sublayers, the velocity decays in a linear fashion throughout the boundary layer. Conversely, in a turbulent boundary layer, owing to the presence of mass interchanges in a direction perpendicular to the surface, energy from the free stream can easily penetrate deep into the boundary layer, so the velocity may only start dropping significantly in the immediate vicinity of the surface, thus leading to a very steep gradient. As a steeper gradient means greater friction drag, it is easy to see why maintaining a laminar boundary layer is highly desirable from an aircraft performance point of view.

In practice, the boundary layers around most surfaces of an aircraft are largely turbulent, though on slender shapes, such as most lifting surfaces, it is possible to create areas of laminar flow, starting at the leading edges. In most cases, as the flow progresses along the top of the lifting surface, this laminar region quickly gives way to a turbulent boundary layer; but, with careful design, this transition can be delayed. More specifically, the streamwise pressure gradient can be engineered in a way that extends the laminar boundary layer further downstream. A *favourable* pressure gradient (i.e. the pressure drops as the flow moves along the surface) will encourage a more gradual velocity decay in the direction perpendicular to the surface, delaying the thickening of the boundary layer, and thus the transition to turbulence. Pressure increasing in a streamwise direction (i.e. featuring an *unfavourable* gradient) will cause increasingly sharper deceleration near the surface, ultimately leading to a reversal of the flow, and thus *separation*.

For Reynolds numbers between 10^5 and 10^7 (calculated with the chord length as reference) the transition will occur very shortly after we enter the adverse pressure gradient zone; that is, after the flow passes the minimum pressure point on the top surface. The logical design measure is therefore to shape the aerofoil in a way that moves this minimum pressure point as far downstream as possible, thus maximizing the extent of the laminar boundary layer over the wing. Caution needs to be exercised here though, as moving the onset of the adverse pressure gradient aft in this way often causes it to be more severe, and thus the transition will be more abrupt too, leading to almost immediate separation. At $Re > 10^7$ the transition point may precede slightly the onset of the adverse pressure gradient. In general, increasing Re will move the transition point forward; at constant Re, increasing incidence will have the same effect (Houghton and Carpenter, 1994).

In addition to the external pressure gradient, the transition of the boundary layer from laminar to turbulent is also influenced by external disturbances, which trigger *Tollmien–Schlichting waves* in the boundary layer – these may ultimately lead to transition. The source of such disturbances can be surface roughness, free-stream turbulence or vibrations, which may significantly degrade the performance of a wing based on a theoretically laminar flow aerofoil. The engineering challenge, therefore, is to design lifting surfaces with *robust* natural laminar flow.

The North American Aviation P-51 Mustang (see Figure 5.2) was the first production aircraft featuring a wing designed for extensive natural laminar flow, but, most likely as a result of the riveting on the upper surface of the wing of the P-51, the laminar flow performance fell short of expectations. There was one serendipitous effect, though: the delayed adverse pressure gradient yielded very good high-speed performance, increasing the drag rise Mach number (Anderson, 2005).

5.2.6 Multi-Element Aerofoils

The concept of multi-element aerofoils is practically synonymous with high-lift-system wing sections. Their importance is summed up by Figure 5.3, a sketch showing typical maximum lift coefficients achievable by a selection of common multi-element aerofoil sections.

Figure 5.2 The North American Aviation (NAA) P-51 Mustang, the first production aircraft to employ an aerofoil designed for laminar flow, the NAA/NACA 45-100 (photograph courtesy of the US Air Force).

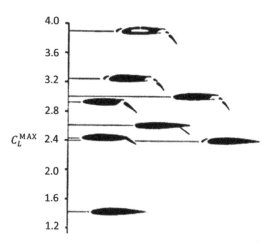

Figure 5.3 Approximate maximum lift coefficient values attainable by various multi-element aerofoil configurations (diagram courtesy of NASA).

It would appear that significant gains in C_L^{MAX} demand increases in the number of aerofoil elements. Alas, this almost always translates into increases in weight, complexity and cost. Here are a few estimates made by Rudolph (1996) in terms of the impact of increasing the number of elements on a trailing edge flap system (the specific weight figures are in terms of the deployed flap area and the cost is expressed in terms of the overall manufacturing cost of the aircraft):

- single-slotted – weight: 8.9 lb/ft^2 (43.4 kg/m^2); part count: 1540; cost: 2.5%;
- double-slotted – weight: 13.2 lb/ft^2 (64.4 kg/m^2); part count: 2430; cost: 5.0%;
- triple-slotted – weight: 15.00 lb/ft^2 (73.2 kg/m^2); part count: 2880; cost: 6.4%.

The conceptual designer's design rationale, however, must go deeper than this. Meredith (1993) points out that a 0.1 increase in lift coefficient at constant angle of attack is equivalent to reducing the approach attitude by about 1°. For a given aft body-to-ground clearance angle, this translates into a shortened landing gear and a weight reduction of over 0.5 t on a large twin-engine transport. Do we win back here enough of the weight sacrifices of a more complicated multi-element section? Once again, there is a hierarchy of objective functions (recall Section 2.7.1), and multi-element section design demands as high a level in this hierarchy as almost any other aspect of the geometry of the airframe.

Parameterization presents a similarly special challenge (see Sobieczky (1998) for a primer), as we are not only having to parameterize the elements themselves, but their relative positions too. Moreover, most flap systems have to be capable of multiple settings corresponding to different points in the operational cycle of the aircraft.

The optimal geometrical design of multi-element lifting surfaces is, in fact, a perfect storm of nearly all the serious challenges we have talked about so far. It features multiple disciplines (flow, structures, reliability, cost), competing objectives with multiple design points (recall the discussion in Section 2.7.2), as well as exceptionally tricky aerodynamics (see the review by van Dam (2002) discussing recent viscous/inviscid coupled and Reynolds-averaged Navier–Stokes-based approaches).

5.2.7   Morphing and Flexible Aerofoils

Aerofoils that can change their shape as commanded by the control systems of the aircraft have been a part of aviation from the very beginning (the Wright Flyer is the most notable early example), but it is a technology that is still at a relatively low readiness level across the aircraft industry.

These aerofoils aim to provide a cheaper, lighter and lower maintenance alternative to the multi-element sections discussed above, and the challenges associated with their geometrical optimization are similar, in particular in terms of the multi-objective, multipoint and multidisciplinary nature of the problem.

An additional challenge is that the parameterization of the wetted surface geometry has to be conceived in conjunction with the (generally also parametric) geometry of the internal structure, as the shape of the former is usually strongly influenced by the latter. A good example is the study by Ursache *et al.* (2006), where the thickness distribution of a post-buckled internal structural member determines the external profile of the wing – the interested reader may wish to consult Barbarino *et al.* (2011) for further examples.

A related technology is that of *flexible aerofoils*, designed to mimic the thin, membrane wings of flying mammals (bats, flying squirrels, etc.). These very thin aerofoils (typical thickness-to-chord ratios are around 2%) have been found to have good stability characteristics in turbulent, low Reynolds number flows (Hu *et al.*, 2008; Null and Shkarayev, 2005).

5.3   Legacy versus Custom-Designed Aerofoils

The title of this section sets out the fundamental dichotomy facing every aircraft designer. Do you use existing, tried and tested legacy sections or do you design your own? The latter method involves building a parametric geometry, whose optimum parameter set is then established through a direct or inverse optimization process under the guidance of some measure of merit (objective function).

In fact, probably the majority of designers view the choice of *aerofoil*(s) defining, say, the main wing of their aircraft, as just that: a *catalogue choice*, not very different from the selection of a fastener or a bearing for a mechanical device. Using legacy *aerofoils* in this way has a number of benefits:

- wind tunnel and computational data are often available on them;
- aerofoils already flying on production aircraft also often offer the benefit of experience on more subtle aspects of their performance, such as handling characteristics (in particular, stall behaviour) and robustness to contamination;
- development costs of aircraft featuring legacy aerofoils can be much lower than those of aircraft with new, bespoke sections.

The strength of the latter argument, however, is on the wane. Improvements in the fidelity and affordability of computational analysis tools are likely to make developing new, custom sections more appealing even on less expensive projects. Of course, legacy aerofoils will continue to play an important role in aircraft design, at the very least as starting points of iterative optimization procedures. These procedures are most often aimed at local fine tuning of a legacy aerofoil for customization to the design task at hand, though legacy sections can serve as 'bases' in a global search too (Sóbester, 2009).

Whether we use a legacy section as it is or as the starting point of an iterative process of local (or, less often, global) improvement, certain precautions are required. We discuss these next.

5.4   Using Legacy Aerofoil Definitions

Many of the legacy aerofoil sections recorded by the literature were defined at some stage of their development through analytical expressions. Such expressions, where they survive, are very valuable, as they can be resampled at will, they can be analysed more readily and sometimes they can be converted to other representations, for instance as required by a CAD engine. Especially useful are those formulations that contain shape control parameters (such as the well-known four-digit NACA formulation, which we will discuss in Section 6.1), allowing their natural and immediate deployment in optimization studies.

However, sadly, the passage of time or intellectual property considerations have resulted in the loss of many of these analytical formulations. In some other cases such algebra never existed, as the shapes have emerged as a result of local reshaping operations driven by wind tunnel tests. The end result of both types of scenario is that we are left with definitions consisting merely of a series of coordinate pairs belonging to points on the aerofoils. Worse still, such coordinate sets are sometimes subject to rounding-off errors or they are insufficiently dense.

Since most practical design and/or analysis exercises demand some sort of analytical description, a *curve fitting* procedure is needed, and this can be somewhat messy.

The key question is: what is the appropriate choice for the *analytical formulation of the fitted curve*? When the type of curve originally used is no longer known, not suitable for current purposes, or never existed, we may (and have to) select a new one.

In Forrester *et al.* (2008) we presented a generic framework for fitting curves (and hypersurfaces) to arbitrary data. This uses a Bayesian reasoning, whereby the parameters of a highly generic curve formulation are estimated by maximizing the likelihood of the training data having come from a postulated fitted model. While some of the basic principles of this methodology could be applied here too, there are good reasons to restrict the range of analytical formulations we may consider fitting to aerofoils. Most importantly, aerofoil fitting requires comparatively limited flexibility (for instance, aerofoils tend not to feature multiple inflection points per surface).

There is a set of constraints usually associated with aerofoil fitting. The issue of smoothness must be considered. C^0 (zeroth order) continuity is an obvious requirement, but aerodynamic analysis considerations generally demand that the curve is differentiable too, at least once; that is, C^1 continuity is sought (sharp leading edge aerofoils, such as that shown in Figure 5.1, are a possible exception, though in practice even these will have a small nose radius).

Another important consideration here is the magnitude of the fluctuations in this first differential. More intuitively, one could consider the related quantity of the unit tangent vector, or, more intuitively still, the *curvature* (which is the rate of change of the former).

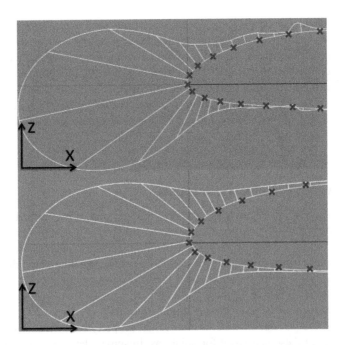

Figure 5.4 Interpolating (top) versus regressing (bottom) with cubic splines – balancing closeness to training data against smoothness of fit. The curvature of the interpolating curve (depicted as a porcupine plot attached to the aerofoil contour) exhibits small 'wobbles', but it goes through each legacy data point (marked by × symbols), while the regression curve has a smooth curvature variation, but it does not stay completely true to the data.

This brings us to a fundamental decision to be made when fitting a curve to legacy data: do we fit an *interpolant* (which passes through each data point) or a *regressor*, which does not necessarily do? Some measure of fluctuation, with a 'period' determined by the spacing between the points, is almost inevitable in the case of interpolating models. This is a reflection of a small-scale 'snaking' in the curve (usually at an amplitude of fractions of a per cent of chord), which is pernicious, because it is not a design feature of the aerofoil, merely an artefact of the model fitting process. It is unwanted noise – an example of such a noisy curvature is shown in the top plot on Figure 5.4.

Conversely, a regressor may not go through the legacy data points, but it could be chosen to minimize the curvature fluctuations (or eliminate them altogether) – as shown on the bottom plot in Figure 5.4 (same data as in the top plot).

Both regression and interpolating models have one more pitfall: they may alter the location of the leading edge point. In purely geometrical terms one may define the leading edge point of an aerofoil as the point farthest away from the trailing edge, measured as the Euclidean distance between the two (in Section 5.1 we saw some aerodynamics-based alternatives). It is possible that this will not coincide with the legacy point with the lowest abscissa value, in which case it will have to be identified through a process of optimization – the elevation angle of a ray emanating from the trailing edge will have to be iterated up and down until the maximum distance is found to within a desired accuracy. The 'optimal' ray will then become the chord. This recipe assumes a sharp trailing edge – a similar method can be devised for finite trailing edges (say, using the midpoint of the trailing edge).

When fitting a curve (of whatever functional form) to a set of legacy coordinates it is also worth considering which geometrical features are most relevant to the aerodynamics of the problem at hand and focus on matching these. Kulfan (2010) suggests the following rules:

- Pressures in subsonic flow around a wing section are governed by surface *curvature*; therefore, matching curvature is important.
- Pressures in supersonic flow around a wing section are governed by surface *slopes*; therefore, matching slopes is important.
- Supersonic far-field wave drag varies with variations in surface *ordinates*; therefore, matching surface ordinates is important.

5.5   Handling Legacy Aerofoils: A Practical Primer

Having introduced some of the fundamental ideas of wing section design, this is a good time to give the reader the opportunity for some experimentation. Later on we will provide the tools needed for the design of new aerofoils; for now, let us take a practical look at some of the basics of handling existing sections.

Perhaps the most common way of storing aerofoil coordinates in human-readable text files is the two-column format used by the University of Illinois at Urbana-Champaign (UIUC) Aerofoil Coordinate Database (sometimes referred to as the *Selig format*, so named after the head of the UIUC Applied Aerodynamics Group). The columns correspond to the x and z coordinates, which are listed starting from the trailing edge, moving forward along the upper surface to the leading edge and then back around the lower surface to the trailing edge.

The MATLAB® function `readfoildata.m` included in the toolkit accompanying this book is designed to parse a coordinate set stored in this format and to break out the x and z vectors corresponding to the two surfaces, returned as `[XUpper, ZUpper, XLower, ZLower]`. An OpenNURBS/Rhino-Python implementation of the same procedure is called by the method `AddAerofoilFromSeligFile` included in the `Aerofoil` class.

The next step is to fit a curve to the imported coordinate set, bearing in mind the discussion in Section 5.4. If the coordinate set is practically the 'mesh' we plan to use as the input to, say, a panel flow solver, we can simply fit a piecewise linear poly-line to the set, but for sparse or locally 'gappy' legacy data a more sophisticated approach is needed – a curve needs to be fitted to the coordinate set that does a better job at filling the gaps than a simple linear model (an easy way of testing the suitability of a model for this type of curve fitting is to hold back some of the available legacy data and fit the model to the rest, and then try to predict that reserve data back using the model – the resulting error then may be compared with, say, the manufacturing errors of the wing construction process).

The toolkit provided with this chapter contains two implementations of legacy aerofoil data curve fitting, each with a different functional formulation.

The MATLAB® version uses the class- and shape-function transformation (CST) of Kulfan (2008). We will discuss CST in ample detail in Chapter 7 – for now, we simply note that the MATLAB® function `findcstcoeff` can be used to fit a CST regression model of a given order to the points.

Listing 5.1 OpenNURBS/Rhino-Python code illustrating the process of fitting a smooth cubic spline to legacy aerofoil data presented in the Selig format.

```
1  #Instantiate the Aerofoil class to set up a generic aerofoil
2  #with these basic parameters: leading edge point, chord length,
3  #angle of rotation in a 'yaw' sense (Rotation) and in a 'pitch'
4  #sense (Twist), as well as the path to the Selig-format file
5  #containing the coordinates
6  Af = Aerofoil(LEPoint,ChordLength, Rotation, Twist, SeligPath)
7
8  #Name of the file containing the aerofoil coordinates and
9  #the variable containing the amount of  regression
10 AerofoilSeligName = 'dae11' #Drela DAE11 low Re aerofoil
11 SmoothingPasses = 2
12
13 #Add aerofoil curve to Rhino document, and retrieve handles to
14 #it and its chord
15 AfCurve,Chrd = Aerofoil.AddAerofoilFromSeligFile(Af, AerofoilSeligName,
16 SmoothingPasses)
```

The OpenNURBS/Rhino-Python implementation fits a spline to the data, to which it can apply successive 'smoothing' passes, essentially increasing the amount of regression, as illustrated in Figure 5.4. Listing 5.1 is an example of this process. First, an object of the `Aerofoil` class is created – at this point we merely have a placeholder for an aerofoil in the appropriate spatial orientation, but no actual geometry yet. For that we call the `AddAerofoilFrom-SeligFile` method included in the `Aerofoil` class – this will actually generate the curve representing the aerofoil, in this case the Drela DAE11 low Reynolds number aerofoil.

5.6   Aerofoil Families versus Parametric Aerofoils

We spoke in Section 5.3 about the distinction between legacy aerofoils and flexible parametric curves suitable for aerofoil geometry parameterization. We also mentioned that aircraft design using the former category essentially treats aerofoils like just another catalogue item, a mindset that does not preclude optimization (the catalogue can be viewed as a discrete-valued search space), but it does not allow the same flexibility as a generic parameterization might.

There is, however, a third category of section definitions, which has some of the advantages of both methodologies: sets of 'family-related' legacy sections. We devote Chapter 6 to these.

6

Families of Legacy Aerofoils

One might loosely define an aerofoil family (or series) as a set of sections developed on the basis of the same design philosophy. Whether there exists an analytical, parametric functional form which all members of the family share, or the family is simply defined through its members, such sets are rather useful from an optimal design perspective. They give the design space a structure that allows us to implement, once more, a 'divide and conquer' approach to design optimization, by breaking the problem down into two or, possibly, three stages:

- the search for the best family – this is the exploratory phase;
- the search for the best member within the family – this is the exploitation phase;
- and an optional stage – we may regard the chosen family member as the starting point of a local fine-tuning process by applying small perturbations to it.

One might substitute 'parametric aerofoils' for 'families' in the scheme above, though here we prefer to reserve the former term for the much more flexible and generic formulations that we will cover in Chapter 7. In fact, the model we begin with here could be most aptly described as a 'parametric legacy aerofoil family', one of the finest and most elegant mathematical models in the history of aeronautical engineering.

6.1 The NACA Four-Digit Section

NACA developed its well-known four-digit lifting surface section formulation in the early 1930s. Perhaps little did they realize that well into the 21st century, although considered somewhat obsolete, the four-digit NACA foils would still be in occasional use in a range of applications, such as light aircraft and low-speed unmanned air vehicles, especially as baseline shapes to start local optimization processes from.

Owing to this lingering popularity, the elegance of its formulation and its (deceptive!) simplicity, we shall discuss this formulation in detail in what follows, using it to illustrate some of the pitfalls it shares with many other aerofoils.

The four-digit series broke new ground in a variety of ways. To begin with, it was a series! That is, it was a family of similar shapes differentiated by a set of parameters (design variables).

Aircraft Aerodynamic Design: Geometry and Optimization, First Edition. András Sóbester and Alexander I J Forrester.
© 2015 John Wiley & Sons, Ltd. Published 2015 by John Wiley & Sons, Ltd.

In fact, as we shall see shortly, these parameters have clear, intuitive effects on the shape, a feature that may account for its enduring popularity into the era of computer-aided, optimal design.

The second major innovation of the NACA four-digit series was that it was the first formulation to split the aerofoil shape into two: a parametric camber line and a parametric (half-) thickness distribution around it. We start the detailed description with this latter element.

6.1.1 A One-Variable Thickness Distribution

'Well-known aerofoils of a certain class including the Göttingen 398 and the Clark Y, which have proved to be efficient, are nearly alike when their camber is removed (mean line straightened) and they are reduced to the same maximum thickness', observed Jacobs et al. (1933) in their seminal paper, which is today seen as the 'birth certificate' of the NACA four-digit aerofoil. Indeed, the generic half-thickness distribution of the NACA four-digit aerofoil was obtained by fitting the polynomial model

$$t(x) = a_0\sqrt{x} + a_1 x + a_2 x^2 + a_3 x^3 + a_4 x^4, \quad x \in [0, 1] \tag{6.1}$$

to some of the key features of the Göttingen 398 and the Clark Y:

- maximum thickness of 10% at 30% chord – that is, $t(0.3) = 0.1$;
- tangent horizontal at maximum thickness point – that is, $dt/dx\,|_{x=0.3} = 0$;
- trailing edge 0.2% thick – that is, $t(1) = 0.002$;
- tangent at trailing edge $dt/dx\,|_{x=1} = -0.234$;
- leading edge shape $t(0.1) = 0.078$.

These yield the coefficients $a_0 = 0.296\,90$, $a_1 = -0.126\,00$, $a_2 = -0.351\,60$, $a_3 = 0.284\,30$ and $a_4 = -0.101\,50$. Inserting these into Equation 6.1 results in an overall aerofoil thickness of 20% chord, so the generic equation for an aerofoil of maximum thickness t_{max} (in units of chord) can be found by dividing through with 0.2 and multiplying by the maximum thickness:

$$t(x) = t_{max}\left(1.4845\sqrt{x} - 0.63x - 1.758x^2 + 1.4215x^3 - 0.5075x^4\right), \quad x \in [0, 1]. \tag{6.2}$$

The thickness distribution is thus governed by a sole variable, t_{max}. Incidentally, this is therefore also true of the leading edge radius, a term that, in the case of this family of aerofoils, is generally used to refer literally to the radius of curvature at the leading edge point.[1] It can be shown (see Ladson et al. (1996)) that thanks to the square-root term in Equation 6.2 this is a finite quantity at $x = 0$ and can be calculated as

$$R_{LE} = \left.\frac{\left[1 + \left(\frac{dt}{dx}\right)^2\right]^{3/2}}{\frac{d^2 t}{dx^2}}\right|_{x=0} \approx 1.1019\,t_{max}^2. \tag{6.3}$$

---

[1] Other aerofoil formulations may use the terms 'leading edge radius' or 'nose radius' to refer to subtly different quantities.

6.1.2   A Two-Variable Camber Curve

The camber curve (or mean curve) $z_{cam}(x)$, $x \in [0, 1]$, of the four-digit NACA foil is composed of two adjoining parabolas: the forward one increasing from zero at the leading edge to the ordinate x_{mc} corresponding to the maximum camber value z_{cam}^{max}, and the aft one decreasing from there to zero at $x = 1$. The derivative of the curve is continuous throughout and it vanishes at the joining point. Formally, the six coefficients $b_0, b_1, \ldots, b_5$ of the generic formulation

$$z_{cam}(x) = \begin{cases} b_0 + b_1 x + b_2 x^2 & x \in [0, x_{mc}] \\ b_3 + b_4 x + b_5 x^2 & x \in (x_{mc}, 1] \end{cases} \tag{6.4}$$

can be computed by imposing

- the end-point conditions $z_{cam}(0) = 0$ and $z_{cam}(1) = 0$, and
- the maximum camber point conditions $dz_{cam}/dx \,|_{x=x_{mc}} = 0$ and $z_{cam}(x_{mc}) = z_{cam}^{max}$,

yielding

$$z_{cam}(x) = \begin{cases} \frac{z_{cam}^{max}}{x_{mc}^2} \left(2 x_{mc} x - x^2 \right), & x \in [0, x_{mc}] \\ \frac{z_{cam}^{max}}{(1-x_{mc})^2} \left(1 - 2 x_{mc} + 2 x_{mc} x - x^2 \right), & x \in (x_{mc}, 1] \end{cases} \tag{6.5}$$

We define $\theta(x)$ as the inclination of the camber curve:

$$\theta(x) = \arctan \frac{dz_{cam}}{dx}, \tag{6.6}$$

where

$$\frac{dz_{cam}}{dx} = \begin{cases} \frac{z_{cam}^{max}}{x_{mc}^2} \left(2 x_{mc} - 2x \right), & x \in [0, x_{mc}] \\ \frac{z_{cam}^{max}}{(1-x_{mc})^2} \left(2 x_{mc} - 2x \right), & x \in (x_{mc}, 1] \end{cases} \tag{6.7}$$

6.1.3   Building the Aerofoil

All the ingredients are now in place for the construction of the NACA four-digit aerofoil. The assembly process involves erecting normals to the camber curve, along which we apply the (half-)thickness distribution t to both sides, obtaining the upper and lower surfaces, (x^u, z^u) and (x^l, z^l):

$$\begin{bmatrix} x^u(x) & z^u(x) \\ x^l(x) & z^l(x) \end{bmatrix} = \begin{bmatrix} x - t(x) \sin \theta(x) & z_{cam}(x) + t(x) \cos \theta(x) \\ x + t(x) \sin \theta(x) & z_{cam}(x) - t(x) \cos \theta(x) \end{bmatrix}, \quad x \in [0, 1]. \tag{6.8}$$

It is worth noting that, because the thickness is not zero at $x = 1$, all aerofoils of the family will have a finite-thickness trailing edge.

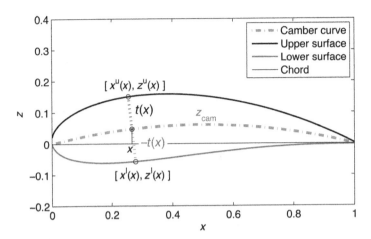

Figure 6.1 Adding/subtracting a NACA one-variable thickness distribution t to/from a NACA two-variable camber curve z_{cam} along normals to the latter will yield the upper/lower surfaces of a NACA four-digit aerofoil.

6.1.4 Nomenclature

The customary designation of NACA four-digit aerofoils follows the template:

$$\text{NACA } \underbrace{100z_{cam}^{max}}_{1\,digit} \; \underbrace{10x_{mc}}_{1\,digit} \; \underbrace{100t_{max}}_{2\,digits} . \tag{6.9}$$

For example, NACA 6521[2] denotes a 21% thick aerofoil with 6% maximum camber at chord station 0.5 (this is, incidentally, the aerofoil depicted in Figure 6.1). The original paper by Jacobs *et al.* (1933) lists 115 possible NACA four-digit aerofoils, resulting from the possible combinations of the following sets of values (all, as before, in units of chord):

$$z_{cam}^{max} \in \{0, 0.02, 0.04, 0.06\}$$
$$x_{mc} \in \{0.2, 0.3, 0.4, 0.5, 0.6\}$$
$$t_{max} \in \{0.06, 0.09, 0.12, 0.15, 0.18, 0.21, 0.25\}.$$

This is what one might call the 'official' range, though there is nothing really special about these aerofoils, other than the availability of experimental aerodynamic data on many of them (Jacobs *et al.* (1933) reported the results of wind tunnel tests on 78 of the 115). However, if there is no need for such known results, one could design lifting surfaces featuring other values – say, a maximum thickness $t_{max} = 0.1$. In fact, the three key numbers could even be treated as continuous design variables in an optimization context, the design search resulting, perhaps,

---

[2] While the name of the institution, NACA, used to be pronounced as individual letters, the 'NACA' prefix in the names of these families of aerofoils is almost universally pronounced as a single word (perhaps mirroring the pronounciation of NACA's successor, NASA, always pronounced as one word).

in $t_{max} = 0.1021$, defining a perfectly valid design (as it actually does in the optimization case study in Chapter 12), whose only drawback is that it cannot be encoded using the template in Equation 6.9.

Some caution does need to be exercised, though, when it comes to choosing sensible limiting values for the three parameters, as some extreme values of camber or extremely far forward maximum camber positions may yield nonsensical shapes. In fact, the NACA five-digit sections were developed to mitigate the latter problem, as we shall see shortly. First, though, we need to examine some practical matters regarding the four-digit family.

6.1.5   A Drawback and Two Fixes

All sensible aerofoils can, in principle, be expressed as a pair of explicit functions, usually separated by the chord line. In other words, one could devise a pair of functions of the form $z^u = f^u(x)$ and $z^l = f^l(x)$ to describe the upper and lower surfaces of any aerofoil shape. This does not mean, however, that all mathematical formulations of aerofoil shapes are of this variety, and Equation 6.8 witnesses to the fact that, for example, the NACA four-digit family is not (owing to the thickness being measured along the normals and not parallel with the z axis). From a computational modelling perspective this can be something of a nuisance – tasks that can be accomplished quite readily on explicit formulations can turn into quite a rigmarole on aerofoils defined implicitly (think, for instance, of determining exactly the vertical space available for a wing spar at a given chord station).

It is tempting to 'cheat' by adding the thickness to the camber curve along the z axis instead of along the normal to the camber curve; this will yield explicit equations describing a sensible-looking aerofoil, but it will not be a NACA four-digit aerofoil (the greater the maximum camber is, the greater the deviations will be).

A better solution is to generate a set of (x^u, z^u) and (x^l, z^l) coordinates as per Equation 6.8 and fit a different, explicit formulation to this 'training data'. This explicit 'surrogate' model could be, for example, the CST (see Section 7.3). A note of caution is in order here, though. There is nothing in Equation 6.8 to suggest that if x is positive (as it must be, by definition), the resulting x^u value would be positive too – negative x^u values *can* occur near the leading edge for nonzero camber cases. Nonetheless, for sensible sampling densities this will only affect the more highly cambered NACA four-digit foils. For example, such an aerofoil is shown in Figure 6.1, which we have cropped very tightly at $x = 0$ to highlight that the [0, 1] box will trim off part of the upper curve near the leading edge (the overhang is about -0.0013). This is not much, and for aerofoils with moderate camber even the second sampled point is likely to be in the positive x region (this is the case for most practical applications – clearly, theoretically, we could generate negative abscissa points by a sufficiently dense sampling on any NACA foil with nonzero camber). In any case, this is something to be aware of when seeking to fit an explicit $z^u = f(x)$-type equation to a set of NACA four-digit aerofoil points – the problem is simply ill-posed for negative x^u values.

6.1.6   The Distribution of Points: Sampling Density Variations and Cusps

Staying with Equation 6.8 and the $x \in [0, 1]$ values, there is another pitfall here. A pertinent question might be: how do we choose these chord stations x to be inserted into Equation 6.8?

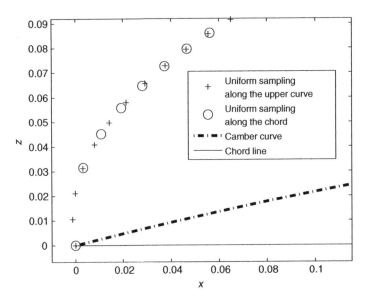

Figure 6.2 Dangers of the naive application of Equation 6.8 – choosing x-coordinates uniformly distributed along the chord may lead to highly nonuniform sampling along the actual aerofoil surfaces, particularly near the leading edge. Shown here is the leading edge region of a NACA 6521 aerofoil sampled both uniformly along x ('o' symbols) and uniformly along the curves themselves ('+' symbols).

If they are uniformly spaced along the chord, the resulting series of (x^u, z^u) and (x^l, z^l) points will not be uniformly spaced along their respective surfaces, especially near the leading edge, where the gradients of z^u and z^l are high – as illustrated in Figure 6.2.

Here is a possible solution. We divide, say, the upper surface into the desired number of segments, assuming zero camber; this is a simple process of numerical integration, as the zero camber means that this curve is explicit – it is simply Equation 6.2. We have implemented this operation in the MATLAB® function `divideexplicit.m`. This will yield a set of abscissas whose distribution takes into account the steep slopes around the leading edge. Applying the required camber (i.e. plugging these abscissas into Equation 6.8) does not spoil significantly the uniformity of the distribution of the points along the aerofoil surface. The MATLAB® function implements this for an arbitrary explicit curve.

Another way of providing a suitable sampling density along the curve is to create a series of abscissas with cosine spacing from zero to a transition point and a linear spacing thereafter, up to one (the transition point corresponding to π on the cos sampling). This method essentially assumes that the nose of the aerofoil can be approximated by a circular segment and produces a distribution that ensures denser fitting of the higher curvature areas – our Python aerofoil code, to be discussed in Section 6.1.8, implements this approach.

A final note of caution in terms of implementations of the four-digit NACA family. We have seen that the two pieces of the camber curve are first-order continuous in their meeting point (which is also the highest point along the z axis; that is, the peak of the camber curve) – the derivatives are equal (zero) on the left and on the right. We discretize this camber curve in some way and we erect normals upon it to generate the lower surface (the length of each line

Listing 6.1   Generating a four-digit NACA aerofoil.

```
1  [Aerofoil,Camber,RLE]=naca4(MaxCamberPercChord,...
2                             MaxCamberLocTenthChord,...
3                             MaxThicknessPercChord,...
4                             Fidelity, N, PlotReq)
```

dictated by the prescribed thickness distribution). On geometries with very high camber one might imagine these orthogonal lines almost intersecting below the maximum camber point if the sampling is sufficiently dense. This means that it is possible to end up with a very slight upward-pointing cusp in the lower surface, which may cause havoc upon meshing or flow analysis. One might eliminate this by deleting the offending cusp point, but a more elegant solution is to fit a smoothing curve to the aerofoil. To this end, we shall use a NURBS curve in the Python aerofoil code accompanying this book (Section 6.1.8) and an explicit CST curve (Section 7.3) in the MATLAB® implementation, which we discuss next.

6.1.7 A MATLAB® Implementation

Our MATLAB® code for the four-digit NACA formulation does ensure uniformity of points along each surface of the aerofoil; however, if the basic, nonuniform formulation is desired (for instance, because it is a little faster to compute, though the difference is only significant if thousands of sections are required), this feature can be turned off. The input parameter `Fidelity` controls this choice, the value `'Low'` will produce an aerofoil sampled uniformly along the chord (as opposed to along the curves themselves). The generic function call is shown in Listing 6.1.

Where the first three parameters come from the template (6.9), `N` is the number of points required on each surface and `PlotReq` determines whether the result is plotted (`PlotReq = 1`) or not (`PlotReq = 0`). The coordinates are returned in `Aerofoil`, a cell array of four $N \times 1$ column vectors containing x^u, z^u, x^l and z^l respectively. The two $N \times 1$ vectors of camber line coordinates (in the cell array `Camber`) and the value of the leading edge radius R_{LE} are also returned. As an example, the command of Listing 6.2 will generate the aerofoil shown in Figure 6.1, a NACA 6521, with 100 points uniformly distributed along each of its two component curves. It will also plot the result.

The four 100×1 surface coordinate vectors will be stored in `Aerofoil` and the two 100×1 camber line coordinate vectors are returned in `Camber`. The leading edge radius is computed as $R_{LE} = 0.0486$. Alternatively, using the generic aerofoil geometry function `aerofoilgen.m`, the same coordinates can be obtained through the call shown in Listing 6.3 – this results in a call to the same `naca4.m` function.

Listing 6.2   Generating the aerofoil shown in Figure 6.1.

```
1  [Aerofoil, Camber, RLE] = naca4(6, 5, 21, 'High', 100, 1)
```

Listing 6.3 Generating the aerofoil shown in Figure 6.1 via `aerofoilgen`.

```
1 Aerofoil = aerofoilgen('NACA4', 'High', 100, 6, 5, 21)
```

Like the other aerofoil generation routines we shall cover in this chapter, `naca4.m` – and thus `aerofoilgen('NACA4',...` too – will produce an aerofoil normalized onto a [0, 1] chord. As we indicated earlier, though, this does not necessarily mean that any point on the aerofoil will have x coordinates in the [0, 1] range – there may be a negative area in the vicinity of the leading edge on the upper surface.

We discussed in the previous section the scenario when an explicitly defined replacement for Equation 6.8 might be desired and we suggested that one might be obtained numerically, say, by fitting a CST explicit function to the coordinate points. We shall cover the CST formulation in much more detail later; for now, we shall just use its MATLAB® implementation to illustrate how one might compute such an explicit surrogate for the NACA four-digit formulation.

Let us assume that we wish to draw the same NACA 6521 aerofoil, complete with a vertical line between its surfaces (say, representing a main spar) at the quarter-chord point. For this we need a means of explicitly finding the z-coordinates of NACA 6521 at $x = 0.25$; Listing 6.4 shows how one might do this.

We deliberately stuck with the rather 'awkward' thick and highly cambered NACA 6521 here to illustrate a comment we made earlier about some of the points near the leading edges of such aerofoils going negative. Mathematically, this means that for some abscissa points (such as the origin) there is no unique ordinate value, so we cannot fit a function of the form $z(x)$ to it. The above piece of code still works though, because `findcstcoeff.m` removes the

Listing 6.4 Generating a NACA 6521 aerofoil in MATLAB® with a vertical line between its surfaces at the quarter-chord point.

```
1  % Generate and plot NACA 6521
2  Aerofoil = aerofoilgen('NACA4', 'High', 100, 6, 5, 21);
3
4  plot([Aerofoil{1}' Aerofoil{3}'], [Aerofoil{2}' Aerofoil{4}'],'+')
5  hold on, axis equal
6
7  % Find the coefficients of the explicit formulation to fit
8  % the coordinates stored in the cell array Aerofoil
9  [CSTCoeffUpper,CSTCoeffLower] = findcstcoeff(Aerofoil,6,7,0);
10
11 % Evaluate and plot the explicit function at an arbitrary x
12 x = 0.25;
13 UpperCurve = cstfoilsurf(x,CSTCoeffUpper);
14 LowerCurve = cstfoilsurf(x,CSTCoeffLower);
15
16 plot([x x],[UpperCurve LowerCurve],'r');
```

Listing 6.5   Generating a four-digit NACA aerofoil as an object of the `Aerofoil` class.

```
1  LEPoint = (0, 0, 0)
2  ChordLength = 1
3  Rotation = 0
4  Twist = 0
5
6  Af = Aerofoil(LEPoint, ChordLength, Rotation, Twist)
7
8  SmoothingPasses = 0
9
10 AfCurve,Chrd = Aerofoil.AddNACA4(Af, 2, 2, 12, SmoothingPasses)
```

(very slightly) negative x points before performing the fit.[3] In other words, it finds the explicit function that fits the original nonexplicit function as closely as possible.

6.1.8   An OpenNURBS/Rhino-Python Implementation

In Section 5.5 we already encountered the OpenNURBS/Rhino-Python class named `Aerofoil` that accompanies this book, which, as the name suggests, provides a means of generating 2D lifting surface sections.

Listing 6.5 illustrates the use of the `Aerofoil` class to generate a NACA2212 aerofoil. First, four key parameters are defined: the location of the leading edge point, the chord length, the rotation angle (this is the 'roll' of the aerofoil) and the twist angle (the 'pitch').

With these basic parameters defined, we can set up a generic aerofoil with these features by instantiating the `Aerofoil` class; this will create the aerofoil object `Af`. Up to this point the aerofoil only exists as a virtual template in a given spatial orientation; now we need to actually draw the aerofoil.

The `AddNACA4` method of the `Aerofoil` class accomplished this and we call it with five arguments: the newly created aerofoil object handle, the four digits of the NACA foil (maximum camber in percentage of chord, the location of the maximum camber in tenths of chord and the maximum thickness of the aerofoil in percentage of chord) as well as the parameter `SmoothingPasses`.

We have discussed aerofoil smoothing in general terms in Section 5.4, and we alluded to reasons why we may need to do this to a four-digit NACA foil in particular in Section 6.1.6. The reader may wish to experiment with the effect of setting `SmoothingPasses` to different values and observing the corresponding curvature diagrams in Rhino (effectively creating a comparison like that shown in Figure 5.4). The greater the value of `SmoothingPasses` is, the smoother the curvature variation will be, but we will also be deviating from the 'true' four-digit NACA profile. Should we use this formulation simply as a starting point of an optimization

---

[3] Recall the discussion in Section 5.1 about the subtleties of what is a sensible location for the leading edge point – sometimes, with parametric aerofoils, such as the NACA four-digit series, varying some of the parameters may cause small ambiguities around whatever way we had chosen to define the leading edge point.

Listing 6.6   The `AddNACA4` method of the `Aerofoil` class.

```
 1  def AddNACA4(self, MaxCamberPercChord, MaxCamberLocTenthChord,
 2  MaxThicknessPercChord, Smoothing=1):
 3
 4          x, z, xu, zu, xl, zl, RLE =
 5          self._NACA4digitPnt
 6              (MaxCamberPercChord,
 7               MaxCamberLocTenthChord,
 8               MaxThicknessPercChord)
 9
10          C = self._fitAerofoiltoPoints(x, z)
11
12          if 'Smoothing' in locals():
13              self.SmoothingIterations = Smoothing
14
15          C, Chrd = self._TransformAerofoil(C)
16
17          return C, Chrd
```

process, this will, of course, have limited importance. However, if the goal is, for instance, the validation of flow analysis code against a known NACA pressure profile, the number of passes should be kept to the minimum the mesher and/or the flow solver is able to cope with (this will probably be zero in most cases).

Listing 6.6 provides a look 'under the hood' of this process. First, a method called `_NACA4digitPnts` is used to generate a set of points representing the specified NACA four-digit aerofoil. For the reasons indicated in Section 6.1.6, this method uses a cos spacing of the abscissa points, implemented in the method `_coslin` (the underscore character precedes the names of methods the user need not call directly – they are internal to the class).

Next, a NURBS curve is fitted to these points via the method `_fitAerofoiltoPoints`. This is then 'post-processed' by the method `_TransformAerofoil`, which applies the specified number of smoothing iterations (using the OpenNURBS/Rhino-Python method `Fair-Curve`), and, most importantly, it applies the offset, scale, 'pitch' and 'yaw' moves specified by `LEPoint`, `ChordLength`, `Rotation` and `Twist` respectively.

6.1.9 Applications

The longevity of the NACA four-digit series has been extraordinary. Nearly eight decades after their inception they are still used occasionally, especially in the design of light aircraft (for instance, the Ikarus C42, a microlight popular in Europe, features a wing based on a NACA 2412) and low-speed unmanned aircraft. They are also to be found sometimes in the role of generators of fairings, tail surfaces and other ancillary elements, as well as in non-aeronautical applications. This should not take anything away, however, from their track record as wing sections, which is quite remarkable, covering a wide range of illustrious aircraft. Without aiming for completeness, some examples include the following: the Douglas DC-2 (NACA 2215 at the root and NACA 2209 at the tip), the Douglas DC-3 (NACA 2215 at the root and

Figure 6.3 The NACA 2213 was used on the Supermarine Spitfire, including on this Type 389 PR XIX, a late, Rolls-Royce Griffon-powered photo-reconnaissance version of the famous World War II fighter. With a top speed approaching 400 knots and a service ceiling in excess of 40 000 feet, the PR XIX was the highest performance Spitfire ever made (photograph by A. Sóbester).

NACA 2206 at the tip), the Scottish Aviation Twin Pioneer (see Figure 6.5), the Supermarine 389 Spitfire PR XIX (see Figure 6.3), the Douglas A1-D Skyraider (Figure 6.4), the Fairchild Republic A-10 Thunderbolt II (see Figure 6.6).

6.2 The NACA Five-Digit Section

6.2.1 A Three-Variable Camber Curve

Two years after their seminal 'NACA four-digit' report, Jacobs and Pinkerton (1935) published a paper describing a variation on their original method, aimed at fixing a flaw in that formulation. We have already hinted at this: the four-digit equations do not work very well when the maximum camber point is located much further forward than about a third of the chord. The fix proposed by Jacobs and Pinkerton (1935) was a new camber curve with maxima between 5% and 25% chord.

Once again, two adjoining curves make up the mean line (camber curve). This time, however, the forward curve is a cubic polynomial, which, upon reaching a transition abscissa m, gives way to a straight line for the aft section of the aerofoil:

$$z_{\text{cam}}(x) = \begin{cases} \frac{1}{6}k_1 \left[x^3 - 3mx^2 + m^2(3-m)x\right], & x \in [0, m] \\ \frac{1}{6}k_1 m^3(1-x), & x \in (m, 1] \end{cases} . \tag{6.10}$$

Figure 6.4 NACA four-digit profiles over Korea and Vietnam. Weighing over 11 t and with a cruise speed of around 170 knots, the Douglas A1-D Skyraider (AD-4) is one of the real 'bruisers' of the history of propeller-driven attack aircraft. The folding wing shows off its NACA 2417 root section (photograph by A. Sóbester).

Figure 6.5 The Scottish Aviation Twin Pioneer, a short take-off and landing transporter designed in the 1950s, had wings featuring NACA 4415 profiles (photograph by A. Sóbester).

Figure 6.6 The Fairchild Republic A-10 Thunderbolt II may be the only jet aircraft with wings based on NACA four-digit sections – a NACA 6716 at the root and a NACA 6713 at the tip (photograph courtesy of the US Air Force).

The above equations result from imposing a second derivative of $k_1(x - m)$ on the polynomial making up the forward section and ensuring that we have second-order continuity at $x = m$. Clearly, the forward curve has to reach its maximum ordinate (the maximum camber) somewhere to the left of m (as opposed to the two-digit camber line formulation, where the maximum camber abscissa coincided with the transition point) – this x_{mc} abscissa can be found by setting the first derivative of the cubic polynomial to zero. This yields two solutions, only one of which is in the range $[0, m]$:

$$x_{mc} = m \left(1 - \sqrt{\frac{m}{3}} \right). \tag{6.11}$$

Finding the transition abscissa m for a given maximum camber abscissa x_{mc}, therefore, amounts to solving the above for m, subject to $m \in [x_{mc}, 1]$. As it happens, for $x_{mc} \in [0.05, 0.25]$ (the range this formulation was designed for), there is always one and only one root that satisfies this constraint.

Two numerical observations are due here. First, Equation 6.11 can be solved by any number of iterative root finders (e.g. the bisection method). There is an alternative though. It can be converted into the cubic equation $m^3 - 3m^2 + 6x_{mc}m - 3x_{mc}^2 = 0$, which can be solved an order of magnitude faster (either using Cardano's equations or by computing the eigenvalues of its companion matrix). At first sight, this is a somewhat questionable manoeuvre, because, theoretically, this cubic equation could have additional roots – but, in this case, it does not. If thousands of sections have to be computed, this sleight of hand can save a significant amount of time, and this is therefore the method adopted in `naca5.m` in the MATLAB® code accompanying this book. A second observation is that the values of m tabulated for a range of x_{mc} values in Jacobs and Pinkerton (1935) are a little inaccurate (these aerofoils date back to two decades before the advent of the electronic computer!), which may explain why there are slight differences between the original aerofoil coordinates and the coordinates calculated as described above.

With this new camber curve in place, the construction of the aerofoil proceeds in exactly the same way as in the four-digit case, using the same thickness distributions – as described by Equation 6.8 (for more details on the calculations the reader may wish to inspect the MATLAB® code provided with this text in `naca5.m` or the notes on geometry modelling by Mason (2009)).

6.2.2   Nomenclature and Implementation

Compared with the elegantly parsimonious and intuitive nomenclature of the four-digit aerofoils, the 'official' five-digit family offers a somewhat less useful shorthand notation. If we wanted a template similar to (6.9), we could postulate one as follows:

$$\text{NACA } \underbrace{\frac{2}{3}c_1^{\text{design}}}_{\text{1 digit}} \times 10 \;\; \underbrace{2x_{mc}}_{\text{1 digit}} \;\; \underbrace{\text{type of camber curve (0/1)}}_{\text{1 digit}} \;\; \underbrace{100t_{max}}_{\text{2 digits}}. \qquad (6.12)$$

The somewhat awkward form of the first digit already hints at the fact that the original family was rather limited. In fact, all (five) of its members were designed for a lift coefficient of 0.3, so their notations all begin with a '2'. Also, they were all 12% thick and covered the maximum camber positions range between 5% and 25% in steps of 5%. They were thus denoted as NACA 21012, NACA 22012, NACA 23012 (see Figure 6.7), NACA 24012 and NACA 25012. The third digit was introduced to accommodate a *reflexed* camber curve series, where the aft straight portion of the curve was replaced with a curve designed to reduce the

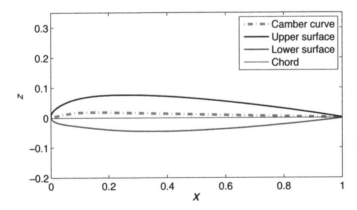

Figure 6.7   NACA 23012 – a 12% thick NACA five-digit aerofoil with a design lift coefficient of 0.3 and maximum camber located at 0.15 units of chord. The transition point between the two parts of the camber curve is at $m = 0.2027$ units of chord – aft of this point the camber curve is a straight line.

Listing 6.7 Generating a NACA five-digit aerofoil using the MATLAB® function naca5.

```
1  [Aerofoil, Camber, RLE] = naca5(DesignLiftCoefficient,...
2                                    MaxCamberLocFracChord,...
3                                    MaxThicknessPercChord,...
4                                    Fidelity, N, PlotReq)
```

Listing 6.8 Generating Figure 6.7.

```
1  [Aerofoil, Camber, RLE] = naca5(0.3, 0.15, 12, 'High', 100, 1 )
```

quarter-chord pitching moment to zero. These were denoted NACA 21112, NACA 22112, NACA 23112, NACA 24112 and NACA 25112 (the reflexed series has seen little subsequent use, so we shall not delve into the details of their construction here).

As in the case of the NACA four-digit family, there is no reason why one could not add new members – once again, the only advantages of the original set of Jacobs and Pinkerton (1935) are the availability of wind tunnel data for its members and the concise notation. If these are not essential requirements, one could, therefore, use the formulation described above as a parametric aerofoil, with c_l^{design}, $x_{\text{mc}} \in [0.05, 0.25]$ and t_{max} as its design variables – this is implemented in naca5.m. The generic call of the function is shown in Listing 6.7.

Here, the last three variables have the same meaning as in naca4.m. Thus, for example, the call of Listing 6.8 will generate the aerofoil (and the plot) shown in Figure 6.7. The same result can be obtained by calling the generic aerofoil function aerofoilgen.m with the arguments shown in Listing 6.9.

To conclude this section on the five-digit NACA formulation, it is worth noting that, although we have been referring to its first design variable as the 'design lift coefficient', the relationship between the geometry of the camber line and this lift coefficient is merely based on a simple analytical model (thin aerofoil theory) and, therefore, it is not necessarily very accurate. Consequently, it is best to treat this 'design lift coefficient' as only slightly more than a surrogate for the amount of camber. In any case, the paper by Jacobs and Pinkerton (1935) contains a wealth of wind tunnel data that should clarify this for the interested reader.

To conclude this foray into some of the important NACA work that laid the foundations of modern wing section design, we note that the 1940s saw the development of further NACA families, notably the 6- and 6A-series (Loftin, 1948; Patterson and Braslow, 1958), designed to maximize the extent of laminar flow over wings. These are families defined through their members; that is, no analytical formulations exist.

Listing 6.9 Generating the same NACA five-digit aerofoil as in Listing 6.8, using aerofoilgen.

```
1  Aerofoil = aerofoilgen('NACA5', 'High', 100, 0.3, 0.15, 12)
```

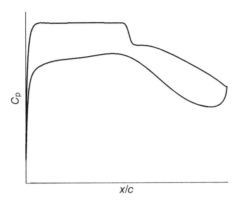

Figure 6.8   Typical 'flat top pressure distribution around an SC(2) supercritical aerofoil.

6.3   The NACA SC Families

In the 1960s, NASA's desire for increasing the drag divergence Mach numbers of transonic aerofoils while retaining acceptable low-speed maximum lift and stall characteristics led to the development of the SC (short for 'supercritical') families of aerofoils (Harris, 1990). They trace their lineage back to the work of Whitcomb and Clark (1965), who noted that a three-quarter chord slot between the upper and lower surfaces of a NACA 64A series aerofoil gave it the ability to operate efficiently at Mach numbers greater than its original critical Mach number. Following on from some of their observations, three SC families were produced, corresponding to three distinct phases of development spanning two decades.

The fundamental design philosophy of the SC aerofoils was to delay drag rise through a reduction in curvature in the middle region of the top surface, limiting flow acceleration and thus reducing the local Mach number. The result is that the severity of the adverse pressure gradient is reduced there, and thus the associated shock is moved aft and weakened.

From an inverse design standpoint, the aim was to create a flat top pressure profile forward of the shock (see Figure 6.8), obtained by balancing the expansion waves emanating from the leading edge, the compression waves resulting from their reflection off the sonic line (separating the subsonic and supersonic flow regions) back onto the surface and a second set of expansion waves associated with their reflection. Geometrically, this was achieved through a large leading edge radius (strong expansion waves) and a flat mid-chord region (reducing the accelerations that would have needed to be overcome by the reflected compression waves (Whitcomb, 1974)). The well-known lower surface aft-end 'cusp' of the SC class of aerofoils is the result of an effort to increase circulation. This helped achieve the target design lift coefficients at low angles of attack (though it also aft-loaded the aerofoil rather aggressively).

6.3.1   SC(2)

While, to the best of our knowledge, no production aircraft ever featured an unmodified NASA SC aerofoil, we discuss them here because this family exemplifies some of the basic principles of transonic aerofoil design. In particular, we shall focus in what follows on the SC(2) series,

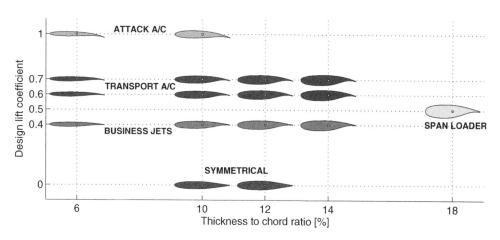

Figure 6.9 NASA SC(2) cross-sections plotted against their thickness-to-chord ratios and design lift coefficients. The labels and colours indicate belonging to an intended class of applications.

as it is a diverse, yet neatly structured family, and many wind tunnel and compute hours have been devoted over the last four decades to understanding its subtleties.

SC(2) comprises 21 aerofoils of different thickness-to-chord ratios and design lift coefficients. A subset of these is shown in Figure 6.9, positioned on the chart according to their design lift coefficients and thickness-to-chord ratios. Their intended main areas of application (as per Harris (1990)) are also shown. Their drag divergence Mach number M_∞^{dd} is usually estimated using an empirical equation attributed to transonic aerodynamics pioneer David Korn (Mason, 2009):

$$M_\infty^{dd} = 0.95 - \frac{t}{c} - \frac{C_L}{10}. \qquad (6.13)$$

The aerofoils are plotted in Figure 6.9 in the orientation in which they were originally introduced; that is, not with their chord horizontal, as it is customary. We follow Harris (1990) by doing this; according to him, 'to simplify comparisons between supercritical aerofoils, it was the custom to present coordinates relative to a common reference line rather than the standard method of defining aerofoils relative to a reference chordline connecting the leading and trailing edges.'

The first two digits of the encoding of each aerofoil represent the design lift coefficient (multiplied by 10), while the third and the fourth digits represent the maximum thickness-to-chord ratio (as a percentage). Thus, for instance, SC(2)-0714 is the 14% thick second series supercritical aerofoil designed for a lift coefficient of 0.7.

The SC(2) series picks up the NACA four-digit idea of defining a camber curve and a thickness distribution around it, with some family members sharing the same thickness distributions – see Figures 6.10 and 6.11 illustrating this.

The MATLAB® function `nearestsc2.m` supplied as part of the toolkit accompanying this book has the design lift coefficient and the thickness-to-chord ratio as its arguments. As the name suggests, it returns the existing SC(2) family member nearest to that pair of values.

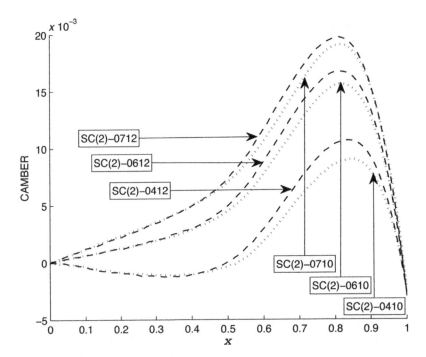

Figure 6.10   The camber curves of six SC(2) aerofoils (axes to different scales).

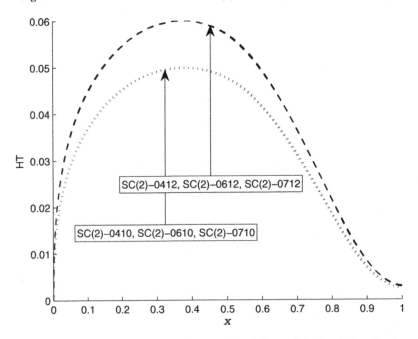

Figure 6.11   The half-thickness (HT) distributions of six SC(2) aerofoils (vertical and horizontal axes are to different scales).

The neat, grid-like arrangement of these aerofoils raises the idea as to whether one might, in some sense, interpolate between them to produce a truly parametric, continuous formulation (in the manner of the NACA four-digit formulation). In particular, the central block of six 'transport aircraft' category aerofoils seems suited to this idea, owing the fact that they fill the space between 10 and 14% thickness and 0.6 and 0.7 lift coefficient densely and uniformly (refer to Figure 6.9). Less uniformly, the 'transport aircraft' and 'business jets' categories fill a broader range, raising the possibility of a more global 'interpolant'.

Of course, 'interpolating' between curves is challenging, especially because we have no analytical expressions for them. An alternative is to encode each aerofoil as a vector of numbers (by 'curve fitting' them using one of the parameterization schemes we shall discuss in Chapter 7) and then build models of the elements of these vectors in terms of design lift coefficient and thickness-to-chord ratio (see Sóbester and Powell (2013) on how to do this and the website accompanying this book for the required tools).

7

Aerofoil Parameterization

In Chapter 6 we reviewed a selection of legacy families of wing section geometries, where 'family' means 'a set of aerofoils created on the basis of the same design philosophy'. In this chapter we tackle the question of how we construct new aerofoils, based on flexible formulations that do not impose, explicitly, a particular aerodynamic design philosophy.

We discuss a number of generic geometry formulations that have the potential to be effective enablers of optimization processes. We have chosen these techniques to represent different balances of the requirements we discussed in Section 2.3 – as with parametric geometries in general, aerofoil descriptions are always trade-offs between satisfying the competing requirements of conciseness, robustness and flexibility.

7.1 Complex Transforms

Complex transforms have, historically, yielded the (mathematically) simplest parametric aerofoil definitions, and thus they are our starting point here. The fundamental idea is this: we generate a (usually simple) geometry and, via a conformal mapping, we transform it into another, in a different space, 'warped' by the complex transformation. This image of the original geometry in the new space is the actual design.

On face value, this is not very practical from a design point of view, because via the complex mapping we may lose the intuitive grasp of the effect of the design variables – these are, after all, the parameters of the original geometry and their effect on the warped geometry might be somewhat difficult to predict. This is, of course, of little relevance to an automated optimizer, but it makes choosing sensible design variable ranges extremely difficult.

This practical inconvenience is often offset to some extent, however, by a rather elegant trait of some mappings, which is that they can also be applied to the potential flow streamlines, not just to the geometry. If these are easy to compute around the simple, initial geometry, this can be a significant bonus feature.

Aircraft Aerodynamic Design: Geometry and Optimization, First Edition. András Sóbester and Alexander I J Forrester.
© 2015 John Wiley & Sons, Ltd. Published 2015 by John Wiley & Sons, Ltd.

7.1.1 The Joukowski Aerofoil

The classic case is the *Joukowski transform*, which maps a circle to an aerofoil shape. The potential flow lines around a rotating cylinder are easy to compute and they are equally easy to map to the flow field around the resulting aerofoil. Here, we are simply concerned with geometry, however, so let us consider how we might construct a parametric aerofoil using the Joukowski transform.

Let $c = x_c + iy_c$ define the centre of a circle of radius R, which we can express as

$$C = R\,e^{it} + c, \quad t \in [0, 2\pi]. \tag{7.1}$$

The corresponding Joukowski aerofoil can simply be expressed as:

$$A = C + \frac{1}{C}. \tag{7.2}$$

In the compact complex algebra notation of MATLAB®, this can be computed as shown in Listing 7.1 (the code to be found in the file `joukowski.m` included in the toolkit accompanying this book is a little longer, as it also includes the capability of plotting the result in the manner of Figure 7.1).

From a parameterization perspective, therefore, we have three variables to play with: the circle radius R and the centre coordinates x_c and y_c (these are not to be confused with the conventional coordinate axes of aircraft geometries – here, we are not in the space of the actual geometry). Figure 7.1 shows an example parameter sweep, illustrating the effect of changing y_c.

While in this particular case y_c appears to be an easy handle for altering the camber of the aerofoil in the mapped space ($y_c = 0$ yields a symmetrical aerofoil), and, similarly, R gives some control of its thickness, the intuitive control ends here. In fact, beyond such simple sweeps, the Joukowski mapping is a somewhat unpredictable geometry generator, which may go some way toward explaining the almost complete lack of practical applications of Joukowski aerofoils on actual aircraft designs. Nonetheless, as an elegant mathematical toy and a milestone in potential flow theory, it deserves its place in a discussion of parametric aerofoil geometries. We shall now a consider a similarly simple, but much more powerful, formulation.

Listing 7.1 Joukowski transform in MATLAB®.

```
1  function Aerofoil = joukowski(xc, yc, R)
2
3  C = xc + 1i*yc;
4  t = (0:0.01:2*pi);
5
6  Circle = R*exp(1i*t) + C;
7
8  Aerofoil = Circle + 1./Circle;
```

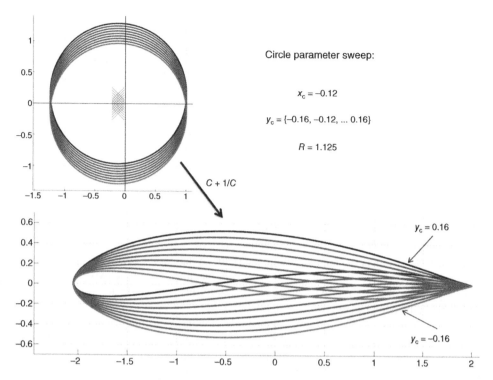

Figure 7.1 One-dimensional parameter sweep on a circle resulting in a family of Joukowski aerofoils of varying camber curves.

7.2 Can a Pair of Ferguson Splines Represent an Aerofoil?

7.2.1 A Simple Parametric Aerofoil

In Section 3.3 we introduced a simple and elegant concept for controlling the shape of a cubic spline: the Ferguson formulation. The simplicity of the Ferguson splines lies in the fact that we can exercise a significant amount of shape control without having to introduce interpolation points between the end points of the curve; we can simply adjust the tensions and the directions of the end-point tangents.

The question therefore arises as to how much geometrical flexibility can be achieved in an aerofoil made up of two Ferguson splines, connected at each end, one for each surface. The concept is simple: the leading edge end-point tangents have to be vertical on both surfaces to ensure first-order continuity, their magnitudes can be used to control the shape of the nose of the aerofoil and the tangents at the trailing edge can be used to control the boat tail angle and the camber, as well as the curvatures of the aft sections of the two surfaces. Figure 7.2 is a sketch of this concept.

The tangent of the upper surface $\mathbf{r}^u(u)$ in $\mathbf{A}$ (at the leading edge) is denoted by $\mathbf{T}_A^{upper}$ and its tangent in $\mathbf{B}$ by $\mathbf{T}_B^{upper}$ – the same nomenclature is used for the lower surface.

The shape of the aerofoil is thus defined by the orientation and the magnitude of the tangent vectors. $\mathbf{T}_A^{upper}$ and $\mathbf{T}_A^{lower}$ will always be pointing vertically downwards and upwards

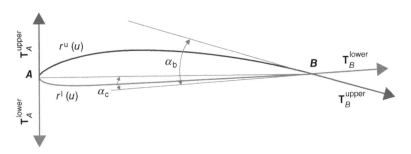

Figure 7.2   A simple parametric aerofoil consisting of two Ferguson splines.

respectively, with their magnitude defining the tension in the spline, thus controlling the 'blunt-ness' of the leading edge. α_c, which might be called the camber angle, defines the orientation of $\mathbf{T}_B^{lower}$, while the boat tail angle α_b determines the orientation of $\mathbf{T}_B^{upper}$. The magnitudes (tangent tensions) of these vectors determine the shape of the middle section of the aerofoil.

We thus have six design variables: the four tangent magnitudes and the two angles – these are the arguments of the MATLAB® implementation of this parametric aerofoil in the code accompanying the book (hermite_aerofoil.m).

For a more general formulation, one may add the option for a finite trailing edge. In this case the curves $\mathbf{r}^u(u)$ and $\mathbf{r}^l(u)$ (see Figure 7.2) will not meet in $\mathbf{B}$, but will have separate end points on a perpendicular raised to the chord in $\mathbf{B}$, at ordinates ε_u and ε_l respectively. This requires no change to the basic formulation, other than replacing $\mathbf{B}(1,0)$ with $\mathbf{B}^u(1,\varepsilon_u)$ and $\mathbf{B}^l(1,\varepsilon_l)$ in the definitions of the two surfaces of the aerofoil.

These could be fixed at certain values, but can also be allowed to function as additional design variables. If this is deemed necessary in an optimization context, a seventh design variable can be added, say, ε, where the top and bottom surfaces are anchored at the trailing edge by $\mathbf{B}^u(1,\varepsilon/2)$ and $\mathbf{B}^l(1,-\varepsilon/2)$ respectively.

A test of the flexibility of an aerofoil parameterization is to attempt to match legacy aerofoils with them. For this type of exercise the search process may be facilitated by using separate ε_u and ε_l trailing edge definition variables – they increase the dimensionality of the problem, but they might make for a less deceptive 'difference landscape' (e.g. when this difference is a mean-squared-error-type measure of the closeness of fit between the parametric and the legacy aerofoils).

So how does the Ferguson spline aerofoil perform in comparative studies of this type? Figure 7.3 shows two examples: a NACA four-digit and a NACA five-digit aerofoil, compared against their best-fit Ferguson spline 'surrogates'. Both the pressure profiles and the geometries themselves show good agreement, indicating that the Ferguson spline-based representation could be used to emulate at least some of the aerofoils from both of these classes – it essentially brings two families under the umbrella of the same formulation (as well as covering the space of other aerofoils based on similar design philosophies). Table 7.1 lists the sets of design variable values that yielded those two approximations.

So how does the Ferguson curve parameterization scheme fare against the criteria listed in Section 2.3? Its *flexibility* is clearly relatively limited – for instance, it would be unable to reproduce aerofoils with multiple inflections or relatively flat portions on either surface, like, for instance, supercritical aerofoils, such as those discussed in Section 6.3 (see Sóbester and Keane (2007) for further comparisons similar to those shown in Figure 7.1). Nevertheless,

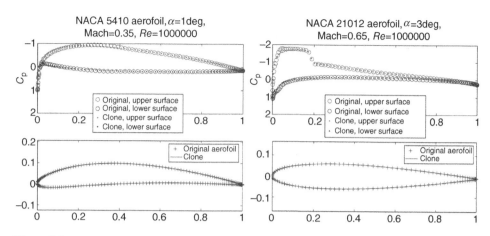

Figure 7.3 Legacy aerofoil geometry matching via the Ferguson-spline-based aerofoil formulation.

it is able to represent a broad range of aerofoils suited to low-speed applications relatively *concisely*, using only six design variables (for a given trailing edge thickness). Finally, its *robustness* is rather good – the only significant pitfall to watch out for here is the possibility of the upper and the lower surfaces intersecting for certain variable combinations. For most design applications it is relatively easy to come up with an appropriate six-dimensional 'box' within which the design variables can move safely – Table 7.2 shows a sensible set of ranges for a typical design study.

7.3 Kulfan's Class- and Shape-Function Transformation

In September 2006, Brenda Kulfan of Boeing Commercial Airplanes presented a paper titled '"Fundamental" parametric geometry representations for aircraft component shapes' (Kulfan and Bussoletti, 2006) in front of a packed room at the 11th AIAA Multidisciplinary Analysis and Optimization Conference held in Portsmouth, VA. Kulfan's *CST*, which is the 'fundamental' parametric geometry representation referred to by the title of that seminal paper, was to become a milestone contribution in geometry modelling and in multidisciplinary design optimization in general. We shall discuss it here in some detail, as it has wide implications to the parameterization of surfaces wetted by flow in general, and its applications span a wide range of aircraft components, design process phases and flow regimes.

Table 7.1 The design variables that generate the Ferguson-spline-based aerofoil description of the best clones of NACA 5410 and NACA 21012.

Aerofoil	$|\mathbf{T}_A^{upper}|$	$|\mathbf{T}_A^{lower}|$	$|\mathbf{T}_B^{upper}|$	$|\mathbf{T}_B^{lower}|$	α_c [deg]	α_b [deg]	ε_1	ε_u
NACA 5410	0.1584	0.1565	2.1241	1.8255	3.8270	11.6983	−0.0032	0.0012
NACA 21012	0.1674	0.2402	2.2482	1.3236	−8.7800	17.2397	−0.0074	−0.0080

Note: these vectors have been obtained via a multi-start minimization of a mean-squared-error metric comparing the two aerofoils at a dense set of sample locations along the chord.

Table 7.2 Typical ranges for the six Ferguson aerofoil definition variables – different (perhaps much narrower) ranges may be better suited to specific design studies.

	$|\mathbf{T}_A^{\text{upper}}|$	$|\mathbf{T}_A^{\text{lower}}|$	$|\mathbf{T}_B^{\text{upper}}|$	$|\mathbf{T}_B^{\text{lower}}|$	α_c (deg)	α_b (deg)
Lower bound	0.1	0.1	0.1	0.1	-15	1
Upper bound	0.4	0.4	2	2	15	30

7.3.1 A Generic Aerofoil

In what follows we shall use a coordinate system whose x-axis is aligned with the chord, with the leading edge point in the origin and the trailing edge point(s) at $x = 1$. We define a universal approximation to any aerofoil in the xOz plane as a pair of explicit curves $\mathcal{A} = [z^u(x, \ldots), z^l(x, \ldots)]$, where $x \in [0, 1]$ and the superscripts 'u' and 'l' distinguish between the upper and the lower surfaces (here and on all the symbols in the following discussion) and the ellipsis indicates that the shapes of the two curves depend on a number of parameters. $\mathcal{A}$ becomes the approximation to a target aerofoil if we determine these parameters such that they minimize some metric of difference (say, mean-squared-error) between $\mathcal{A}$ and the target.

Perhaps the best way to approach the class–shape transformation of Kulfan (2008) is to consider it as a universal approximation formulation for upper and lower aerofoil surfaces. We shall see that it is indeed able to (a) approximate practically any aerofoil (flexibility) and (b) require a relatively small number of design variables to do so with high accuracy (conciseness) – see Kulfan (2006) for the empirical and analytical underpinning of this.

Let the generic aerofoil be defined as

$$\mathcal{A}(\mathbf{V}) = \mathcal{A}\left[x, v_0^u, v_1^u, \ldots, v_{n_{\text{BP}}^u}^u, z_{\text{TE}}^u, v_{\text{LE}}^u, v_0^l, v_1^l, \ldots, v_{n_{\text{BP}}^l}^l, z_{\text{TE}}^l, v_{\text{LE}}^l\right]$$

$$= \left[z^u\left(x, v_0^u, v_1^u, \ldots, v_{n_{\text{BP}}^u}^u, z_{\text{TE}}^u, v_{\text{LE}}^u\right), z^l\left(x, v_0^l, v_1^l, \ldots, v_{n_{\text{BP}}^l}^l, z_{\text{TE}}^l, v_{\text{LE}}^l\right)\right], \qquad (7.3)$$

where n_{BP}^u and n_{BP}^l denote the orders of sets of Bernstein polynomials[1] that control the shapes of the two curves that make up the aerofoil. The upper surface of the aerofoil is defined as

$$z^u\left(x, v_0^u, v_1^u, \ldots, v_{n_{\text{BP}}^u}^u, z_{\text{TE}}^u, v_{\text{LE}}^u\right) = \underbrace{\sqrt{x}(1 - x)}_{\text{class function}} \underbrace{\sum_{r=0}^{n_{\text{BP}}^u} v_r^u C_{n_{\text{BP}}^u}^r x^r (1 - x)^{n_{\text{BP}}^u - r}}_{\text{scaled Bernstein partition of unity}}$$

$$+ \underbrace{z_{\text{TE}}^u x}_{\substack{\text{trailing edge} \\ \text{thickness term}}} + \underbrace{x\sqrt{1 - x}\, v_{\text{LE}}^u (1 - x)^{n_{\text{BP}}^u}}_{\substack{\text{supplementary leading edge} \\ \text{shaping term}}}, \qquad (7.4)$$

---

[1] We have already encountered these as the fundamental building blocks of Bézier curves – see Chapter 3.

where $C^r_{n^u_{BP}} = \dfrac{n^u_{BP}!}{r!\left(n^u_{BP}-r\right)!}$. A curve built upon the same template defines the lower surface:

$$z^l\left(x,v^l_0,v^l_1,\ldots,v^l_{n^l_{BP}},z^l_{TE},v^l_{TE}\right) = \underbrace{\sqrt{x(1-x)}}_{\text{class function}} \underbrace{\sum_{r=0}^{n^l_{BP}} v^l_r C^r_{n^l_{BP}} x^r(1-x)^{n^l_{BP}-r}}_{\text{scaled Bernstein partition of unity}}$$

$$+ \underbrace{z^l_{TE}x}_{\substack{\text{trailing edge}\\\text{thickness term}}} + \underbrace{x\sqrt{1-x}\,v^l_{LE}(1-x)^{n^l_{BP}}}_{\substack{\text{supplementary leading edge}\\\text{shaping term}}}. \qquad (7.5)$$

Taking a general glance at these two equations it is easy to see that at their heart lies a basis-function-type approximation of a target function z: we are expressing the shape of a surface (in this case, the two surfaces of the aerofoil) as the linear combination of a set of simple functions; that is, Bernstein polynomials. The alternative term *partitions of unity* is sometimes used for the Bernstein polynomials to indicate that the sum of all such polynomials of a given order is one.

Looking more closely, as highlighted by the curly braces under the equations, the lower level structure of the transformation has three main components:

- A product of a linear combination of the set of Bernstein polynomials and the so-called *class function*. The latter determines the general family the transformed geometrical entity belongs to; the class function commonly used for geometries of the aerofoil class is $\sqrt{x(1-x)}$. The origins of this are simple: the square root ensures that we can create a rounded nose, as well as imposing an intersection with the abscissa at the origin, while the $1-x$ term enforces an intersection at $x=1$. Figure 7.4 is an illustration of the shapes generated by this first component, broken down into the terms corresponding to the individual Bernstein polynomials. Referring back to the last item on the pre-optimization checklist of Section 2.2.1, it may be noted that each polynomial affects the entire geometry (the entire aerofoil surface in this case), thus giving each variable a global effect. That said, the influence of each term decays relatively quickly; so, for example, in the case illustrated in Figure 7.4, the effect of the zeroth-order shape function term on the aft part of the aerofoil is, in practical terms, negligible.
- A trailing edge thickness term. This is basically a 'wedge' of the same length as the aerofoil. It can be used to elevate the trailing edge to a specified ordinate z_{TE}. Alone, or combined with a similar term included in the other surface of the aerofoil, this allows the construction of a finite-thickness trailing edge.
- The 'supplementary leading edge shape term' – a term designed to increase flexibility in the description of the leading edge by introducing a linear term at the nose. While not essential – the first two terms give adequate approximations in the majority of cases – some highly cambered and certain symmetrical aerofoils benefit from the additional flexibility (Kulfan, 2010). The formulation of this component can be regarded in a variety of ways. In Equations 7.4 and 7.5 we have broken it down to make it obvious that it is, in fact, a mirror image around $x=0.5$ of the last term of the main (top) component the formulation.

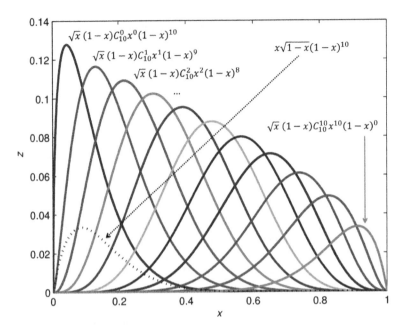

Figure 7.4 The terms of Equation 7.4 (or 7.5) for $n_{\text{BP}} = 10$ and all coefficients v set to one.

Regarded another way, substituting $1 - x$ for x in the $r = n_{\text{BP}}$ term of the main component will yield this term, as made clear by its dotted line plot in Figure 7.4. Equally, one may consider this term as just another shape function to be included into the first line of Equations 7.4 and 7.5: $x^{0.5}(1 - x)^{n_{\text{BP}} - 0.5}$ (to be multiplied with the general class function $\sqrt{x(1 - x)}$, just like the Bernstein polynomial terms).

So what do the basis functions of Figure 7.4 look like when scaled to yield the best approximation of a 'real' aerofoil? Figure 7.5 illustrates the way in which the upper and lower surfaces of a supercritical aerofoil can be built up from the linear combination of four and three Bernstein polynomials respectively.

It highlights the fact that, while each Bernstein polynomial affects the entire span (from 0 to 1), the lower order polynomials (as well as the mirrored version of the highest order polynomial) generally control the regions near the nose, while those of higher orders describe the aft part.

Let us now consider the process whereby we can compute the set of coefficients that represent an existing target wing section – as the structure of Equations 7.4 and 7.5 suggests, this is a simple linear algebra operation.

7.3.2 Transforming a Legacy Aerofoil

Approximating an arbitrary smooth aerofoil with the CST formulation amounts to selecting suitable polynomial orders n_{BP}^{l} and n_{BP}^{u} (as a function of the desired approximation accuracy,

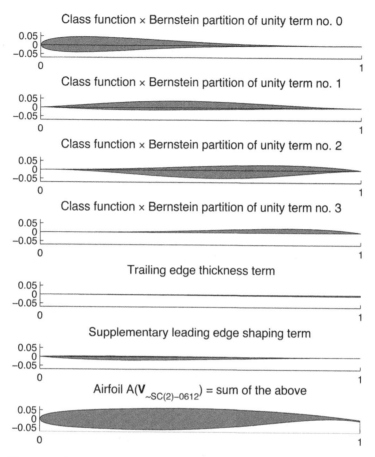

Figure 7.5   The terms of the Kulfan transform of the aerofoil SC(2)-0612, denoted $A(\mathbf{V}_{\sim SC(2)\text{-}0612})$. Note that only the lower surface has a term no. 3 (a positive one in this case), because we used four terms to describe it $\left(n_{BP}^l = 3\right)$ and only three for the upper surface $\left(n_{BP}^u = 2\right)$.

more on which presently) and then finding the vectors

$$n_{BP}^u + 2 \text{ design variables}$$
$$\text{to define upper surface}$$

$$\mathbf{v}^u = \left\{ \overbrace{v_0^u, v_1^u, \dots, v_{n_{BP}^u}^u} , v_{LE}^u \right\}^T \tag{7.6}$$

and

$$n_{BP}^l + 2 \text{ design variables}$$
$$\text{to define lower surface}$$

$$\mathbf{v}^l = \left\{ \overbrace{v_0^l, v_1^l, \dots, v_{n_{BP}^l}^l} , v_{LE}^l \right\}^T . \tag{7.7}$$

that optimize the approximation of the formulation of Equations 7.4 and 7.5 in a least-squares sense, with respect to the target aerofoil. Technically, z_{TE}^u and z_{TE}^l are also variables, but we can simply set them to the trailing edge abscissa values of the target.

Let us consider, say, the upper surface of a target aerofoil, given as a list of n_T^u coordinate pairs $\{(x_{Ti}^u, z_{Ti}^u)|i = \overline{1, n_T^u}\}$. We can exploit the linearity (in terms of the design variables) of the Kulfan transform by rearranging Equation 7.4 in matrix form, equating each of these target points with their approximations:

$$\mathbf{B}^u \cdot \mathbf{v}^u = \mathbf{z}^u, \tag{7.8}$$

where

$$\mathbf{z}^u = \left\{ z_{T1}^u - z_{TE}^u x_{T1}^u, z_{T2}^u - z_{TE}^u x_{T2}^u, \cdots, z_{Tn_T^u}^u - z_{TE}^u x_{Tn_T^u}^u \right\}^T \tag{7.9}$$

and $\mathbf{B}^u$ is an $n_T^u \times (n_{BP}^u + 2)$ matrix of the class–shape function transformation terms, comprising the Bernstein polynomials

$$B_{p,q} = \sqrt{x_{Tp}^u} \left(1 - x_{Tp}^u\right) C_{n_{BP}^u}^{q-1} x_{Tp}^{u\ q-1} \left(1 - x_{Tp}^u\right)^{n_{BP}^u - q + 1}, \quad p = \overline{1, n_T^u}, q = \overline{1, n_{BP}^u + 1} \tag{7.10}$$

and the leading edge shaping terms

$$B_{p,n_{BP}+2} = x_{Tp}^u \sqrt{\left(1 - x_{Tp}^u\right)} \left(1 - x_{Tp}^u\right)^{n_{BP}^u}, \quad p = \overline{1, n_T^u}. \tag{7.11}$$

Note the absence of the trailing edge thickness term from the matrix; this is because we have subtracted the trailing edge 'wedge' from the target coordinate set (Equation 7.9), and we are therefore dealing with a sharp trailing edge (we shall return to this in Section 7.3.5). Computing

$$\mathbf{v}^u = \mathbf{B}^{u+} \mathbf{z}^u, \tag{7.12}$$

where

$$\mathbf{B}^{u+} = (\mathbf{B}^{uT} \mathbf{B}^u)^{-1} \mathbf{B}^{uT} \tag{7.13}$$

is the Moore–Penrose pseudo-inverse of $\mathbf{B}^u$, will now yield the set of coefficients that corresponds to a least-squares fit through the points of the target aerofoil. Naturally, the same procedure can be repeated for the lower surface.

7.3.3   Approximation Accuracy

A key variable of the Kulfan transformation is the number of polynomial basis functions representing each aerofoil surface; that is, n_{BP}^u and n_{BP}^l. Clearly, this is a control directly connected to the flexibility of the scheme; and it is our principal means of controlling the

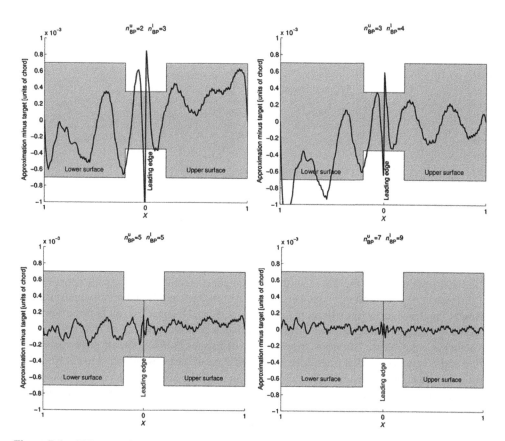

Figure 7.6 CST approximation accuracy at various polynomial orders with the SC(2)-0612 supercritical aerofoil as the target shape.

flexibility versus conciseness trade-off. More parsimonious descriptions – that is, low values of n_{BP}^{u} and n_{BP}^{l} – are desirable from an optimization point of view, but Occam's razor must be set against the range of aerofoil shapes we wish to cover.

One way of gauging the flexibility afforded by different polynomial orders is to select a legacy aerofoil represented by a set of coordinates and apply Kulfan transforms of various orders to it, assessing the accuracy in terms of the deviations between the least-squares regression curves represented by the transforms (obtained via Equation 7.12) and the coordinate points, which represent the 'training data'. Consider Figure 7.6, which depicts a study of this type for the case of the SC(2)-0612 supercritical aerofoil.

Kulfan (2008) sets an approximation accuracy target based on the manufacturing accuracy of a typical wind tunnel model – the shaded regions on the plots in Figure 7.6 represent the corresponding error band (note the tighter tolerances near the leading edge).

The first general observation we can draw from the four tiles of Figure 7.6 is that the double inflection typical of the lower surface of the SC(2) series of aerofoils makes it costlier (in terms of polynomial terms required) to represent than the upper surface. Consider, for instance, the $n_{BP}^{u} = 5$ and $n_{BP}^{l} = 5$ case (bottom left): the amplitude of the deviations is clearly greater on

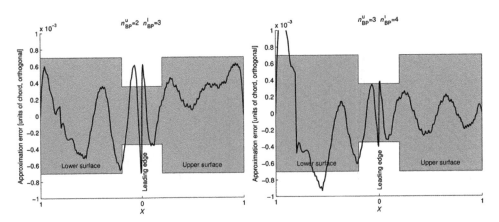

Figure 7.7 The top two tiles of Figure 7.6 repeated using the orthogonal distance metric.

the left (corresponding to the lower surface). A slight difference still persists between the two surfaces at $n_{\mathrm{BP}}^{\mathrm{u}} = 7$ and $n_{\mathrm{BP}}^{\mathrm{l}} = 9$; here, in spite of the two additional polynomials, the aft section of the lower surface is still more difficult to resolve. Nonetheless, on both surfaces we are down to maximum deviations of around 0.0001 of a chord length.

Another conclusion we can draw from this set of images is that some of the errors are a feature of this way of comparing aerofoils, rather than one of the transformation itself. Note the increase in deviations in the immediate vicinity of the leading edge in both cases, particularly noticeable at the lower orders (top tiles). This is due to the fact that the derivative of the aerofoil is approaching infinity here and, therefore, the simple vertical subtraction of the target points from the corresponding abscissa point of the Kulfan transform is a rather unstable calculation. A better alternative is to compute the distance between the target point and the approximation curve, measured along a normal to the curve, erected such that it goes through the target point. Figure 7.7 repeats the top two tiles of Figure 7.6 using this new difference metric.

We are dealing with absolute distances here (instead of the positive or negative deviations arising from the simple subtractions along the z coordinate), but, for the purposes of these plots, we have adopted the following sign convention. On the upper surface of the aerofoil, if the target point is *inside the CST aerofoil* we multiply the normal distance by -1. On the lower surface we multiply distances by -1 if they are *outside* the CST aerofoil. This is a somewhat arbitrary convention, but its consistent application ensures a better understanding of the behaviour of the approximation error (e.g. does the line of target points 'snake' around the approximation?).

Comparing the top two tiles of Figure 7.6 with Figure 7.7, the more predictable behaviour of the orthogonal distance metric in the vicinity of the leading edge becomes apparent (in spite of the low-order approximations), so this is the measure we shall use for the next set of comparisons, depicted in Figure 7.8.

The four panels of the figure correspond to four different aerofoils. In each case we have identified the lowest Bernstein polynomial orders that will get the approximation accuracy within the wind tunnel model tolerance bands.

Clearly, different aerofoils feature different complexities in terms of their surface shapes, and this is reflected in the range of minimum polynomial orders. For example, while the Selig

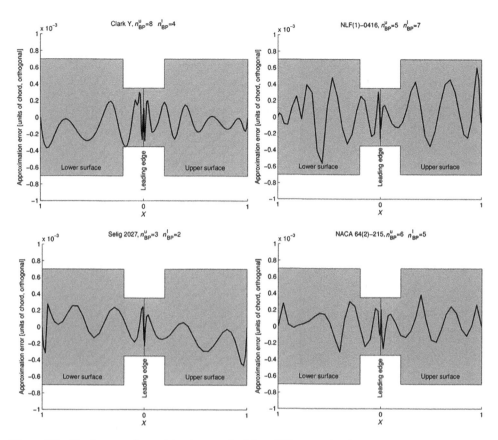

Figure 7.8 The lowest polynomial orders required for a CST approximation to capture a legacy aerofoil within wind tunnel model error varies between different aerofoils – here is a selection.

2027 low Reynolds number aerofoil is captured by upper and lower surface orders of three and two respectively, the natural laminar flow aerofoil NLF(1)-0416 requires five and seven respectively.

7.3.4   The Kulfan Transform as a Filter

One of the features shared by the error curves of Figures 7.6, 7.7 and 7.8 is that they all appear to feature two distinct components: a finer scale random 'noise' component, superimposed on larger scale fluctuations. These mirror two sources of errors, between which we must make a clear distinction:

• Deviations resulting from the approximation of that particular order being 'too stiff' to capture the shape of the target aerofoil. The amplitude of these deviations can be reduced in a predictable fashion by the addition of a few more polynomial terms. This is the source of the large-scale deviations.

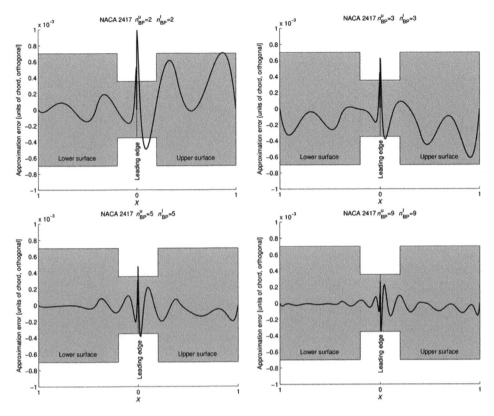

Figure 7.9 Approximation errors of Kulfan curves of various orders fitted to an analytically constructed NACA four-digit aerofoil (NACA 2417).

- Numerical error inherent in the target coordinate set. This is a very common phenomenon with legacy coordinate sets, which may have resulted, for instance, from taking measurements (with their inherent error) off wind tunnel models. Additionally, over time and over multiple transcriptions, further random numerical corruption of these data sets may have occurred. This is the source of the small-scale random fluctuations.

A simple way of checking the above hypothesis is to apply the Kulfan transform to an analytically calculated aerofoil shape – this should be completely free of the fine-scale noise. Consider Figure 7.9, a series of CST model fit errors on a NACA 2417 aerofoil generated via the process described in Section 6.1. Clearly, there is no discernible random error (after all, we are fitting pairs of polynomials!) – the only deviations we can see are those due to the discrepancy between the functional form of the Kulfan transform and that of the NACA four-digit profile. The 'wavelength' of these fluctuations is inversely proportional to the number of Bernstein polynomial terms – the two curves 'snake' around each other at all polynomial orders, but the greater the number of these terms, the more able the CST model is to 'keep up'.

All this amounts to the observation that, beyond its place in optimization, the CST is an excellent tool for smoothing legacy coordinate sets, by fitting a 'clean' model to noisy point sequences, a model, which has the added advantage of being a symbolic, explicit function of x. The computational toolkit provided with this book includes a MATLAB® function designed for just this purpose: findcstcoeff, more on which in the next section.

7.3.5 Computational Implementation

The first step in the construction of a CST curve is the assembly of the matrix **B** of Equation 7.12. In describing the mathematical formulation of this linear system of equations we hinted that we have two options here. Given a set of legacy coordinates we are fitting a Kulfan approximation to, if the aerofoil they represent features a finite-thickness trailing edge, we can either close the trailing edge first (by subtracting a wedge formed of the leading edge and the upper and lower trailing edges) and then build **B** omitting the trailing edge term (this is what we did in Section 7.3.2) or we can build a **B** that does include a trailing edge term and let the least-squares solution process find an approximation that best fits the original set of legacy points. We chose the latter, more generic method for our computational implementation, as illustrated by the MATLAB® snippet of Listing 7.2.

There is little difference between the two approaches, except that the latter does not guarantee that the approximation will pass *exactly* through the legacy coordinate set of trailing edge

Listing 7.2   Generating the matrix **B** for a given Bernstein polynomial order n_{BP}.

```
 1  function B = cstmatrix(nBP,x)
 2
 3  N1 = 0.5;
 4  N2 =   1;
 5
 6  x = x(:);
 7
 8  B = zeros(length(x),nBP+3);
 9
10  % Class function
11  C = x.^N1.*(1-x).^N2;
12
13  for r=0:nBP
14  % Shape function
15      S = nchoosek(nBP,r)*x.^r.*(1-x).^m(nBP-r);
16      B(:,r+1) = C.*S;
17  end
18
19  % Trailing edge thickness term(s)
20  B(:,nBP+2) = x;
21
22  % Leading edge shaping term(s)
23  B(:,nBP+3) = x.^N2.*(1-x).^N1.*(1-x).^nBP;
```

Listing 7.3 Code snippet illustrating the computation of CST coefficients for a given set of coordinates. The complete function can be found under the name `findcstcoeff` in the website accompanying the book.

```
1
2   % Remove trailing edge 'wedge' term from target aerofoil
3   ZTEUpper = ZTargetUpper0(end);
4   ZTELower = ZTargetLower0(end);
5   ZTargetUpper = ZTargetUpper0 - XTargetUpper*ZTEUpper;
6   ZTargetLower = ZTargetLower0 - XTargetLower*ZTELower;
7
8   % Compute coefficients for the upper surface
9   B = cstmatrix(nBPUpper,XTargetUpper');
10  CSTCoefficientsUpper = B\ZTargetUpper(:);
11
12  % Insert the trailing edge 'wedge' back
13  CSTCoefficientsUpper(end-1) = ZTEUpper;
14
15  % Compute coefficients for the lower surface
16  B = cstmatrix(nBPLower,XTargetLower');
17  CSTCoefficientsLower = B\ZTargetLower(:);
18
19  % Insert the trailing edge 'wedge' back
20  CSTCoefficientsLower(end-1) = ZTELower;
```

points – this is something we may or may not care about. Of course, the function of Listing 7.2 still works if we call it for a sharp trailing edge, whether this is the result of that aerofoil featuring one of these *ab initio* or because, as in Listing 7.3, we have removed the wedge prior to calling it.

The function `findcstcoeff` accomplishes the task of 'encoding' a legacy coordinate set into a set of CST coefficients. The code in Listing 7.3 is the core of this. It begins by closing the trailing edge of the target coordinate set, which is contained in the vectors `XTargetUpper`, `ZTargetUpper0`, `XTargetLower` and `ZTargetLower0`. Then the MATLAB® 'slash' (/) operator is used to solve Equation 7.12 and the trailing edge wedge is reinserted.

The computation of the CST curve for another set of abscissas is now a simple matter of multiplying the matrix **B** obtained via the function of Listing 7.2 with the set of coefficients computed as shown in Listing 7.3 – this is basically the computational implementation of the matrix form of Equations 7.4 and 7.5, built upon a vector of x coordinates.

A final note on an implementation issue is due here. In Figure 6.2 we have already highlighted this type of pitfall in the context of the NACA series of aerofoils: if the abscissas we would like to compute the values of the CST approximation for are uniformly distributed, the ordinates will *not* be uniformly distributed along the curve. The function `uniformcstfoilsurf.m` included in the toolkit accompanying this book implements a possible solution: it uses an optimization algorithm to reposition the points along their x coordinates such that they split the curve into equal sections. If other distributions are desired (e.g. density proportional to curvature) the reader may wish to modify `uniformcstfoilsurf.m` to achieve that. A cosine distribution of input abscissas near the leading edge is another possible approach.

7.3.6 Class- and Shape-Function Transformation in Optimization: Global versus Local Search

We have already considered CST as an aerofoil-specific curve-fitting tool, which can be used to capture the shape of any legacy set of aerofoil coordinates as a vector of $n_{BP}^u + 3 + n_{BP}^l + 3$ numbers, but what if we were to regard these numbers as design variables and the Kulfan transformation as a parametric aerofoil geometry?

In terms of the criteria formulated in Section 2.3, the selection of n_{BP}^u, n_{BP}^l and the ranges of the coefficients (the v values from Equations 7.4 and 7.5) will be the decisive choices, but let us take a step back and consider the broader question: what is the scope of the envisaged optimization process? Figure 7.10 provides a hint of some of the difficulties here.

Take the four examples of aerofoil families depicted in the figure. The first two cover a relatively wide spectrum of shapes – we are staying within the same family in each case, but there are some large-scale changes. Correspondingly large variations in their Kulfan coefficients reflect this. These ranges are akin to the design space of an optimization process with a relatively broad scope – certainly broad compared with those suggested by the other two examples, the NACA five-digit set and the supercritical aerofoils shown on the bottom tiles of the figure. The latter is a good example of the sorts of design variable ranges one might use in a local optimization study.

The ranges shown in Figure 7.11 are more representative of the other end of the optimization spectrum. The two black lines connect the limits of the ranges swept by the Kulfan variables

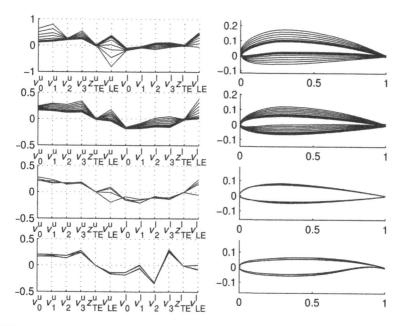

Figure 7.10 Sets of aerofoils (right) and the corresponding sets of third-order Kulfan transform coefficients (left). Top row: NACA 6308 to NACA 6325 (maintaining camber and camber location); second row: NACA 0012 to NACA 6312 (maintaining thickness and maximum camber location); third row: NACA 21012 to NACA 25012; fourth row: SC(2)-0710, −0712, −0610, −0612.

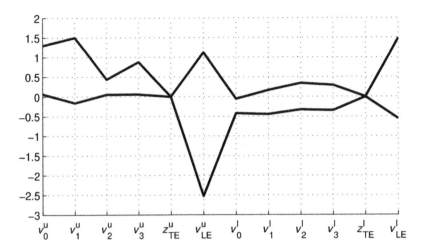

Figure 7.11   Upper and lower bounds on the Kulfan variables of a third order transformation applied to the entire NACA four-digit set.

while covering the entire range of NACA four-digit aerofoils (to be more precise, cambers ranging from 0 to 9%, maximum camber locations between two- and five-tenths of the chord, and thicknesses between 5 and 30% of chord).

All of the above is by way of illustrating the following key point. *The best feature of the Kulfan transformation is its flexibility.* In fact, Equations 7.4 and 7.5 can be considered as quasi-universal approximators of explicit (i.e. of the form $z = f(x)$) aerofoil shapes. From a practical point of view, they can be used to achieve any sensible aerofoil shape to within a specified accuracy, *given a sufficient number of polynomial terms*. It is thus a scheme of variable flexibility too, with the polynomial order in the role of flexibility control parameter.

However, careless use may render optimization searches doomed by the curse of dimensionality. There are circumstances where increasing n^u_{BP} and n^l_{BP} may have an affordable impact on the cost of design searches; for instance, when the objective and/or its derivatives are very cheap or the search (be it direct or indirect) is relatively local. In global searches, especially those involving wings with multiple sections undergoing optimization at the same time, the Kulfan transformation should not be viewed as a *panacea* for all the evils of high-dimensionality design spaces. Nor is it, however, as susceptible to those as, say, a naive parameterization scheme consisting of a NURBS polygon of numerous vertices that are free to move along the x- and z-axes – after all, the class function ensures that, for sensible coefficient ranges, the search will always stay within the realm of 'aerofoil-like' shapes.

7.3.7   Capturing the Shared Features of a Family of Aerofoils

Seeking the best balance between flexibility and search space size in a CST-based optimization process raises the question of how the search could be limited to a certain category or family of aerofoils. To put this in the form of a specific example, how do we choose the ranges of the Kulfan coefficients in an application featuring a cruise Mach number of 0.85, without polluting

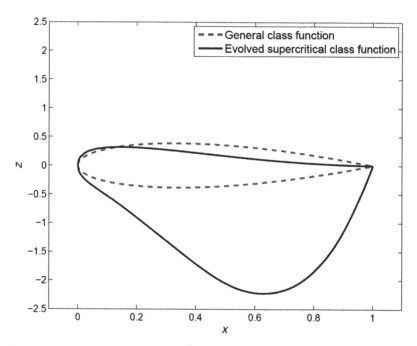

Figure 7.12 The general class function ($\sqrt{x}(1-x)$) and a class function evolved specifically for supercritical aerofoils Reproduced from Powell (2012).

the design space with an infinity of shapes that look nothing like a supercritical aerofoil? This is, of course, a difficult question to answer in the context of any super-flexible scheme; it is by no means a CST-specific problem (e.g. see Juhász and Hoffmann (2004) on tackling the problem for B-splines). However, CST does have the potential to offer some neat ways around the problem.

For example, its elegant modularity offers the tantalizing possibility of replacing the standard aerofoil class function element of the transform with one specifically designed for a family of aerofoils. Take the class function shown in Figure 7.12 as an example, which was developed by Powell (2012) specifically for supercritical aerofoils. The shape of this class function is not particularly intuitive, but this is a relatively unimportant trait in an automated optimization process.

As shown by the comparison depicted in Figure 7.13, when attempting to fit CST aerofoils to the family of SC(2) supercritical aerofoils (which formed the basis of the development of this new class function), for the same polynomial order the approximation errors are much lower than when using the standard class function.

Another avenue is to insert an interim mapping between the optimizer and the Kulfan variables, which links the latter to a higher tier of variables via analytical expressions derived from fitting CST models to large numbers of aerofoils representative of the class relevant to the design case at hand (Sóbester and Powell, 2013). Straathof (2012) discusses further approaches exploiting the elegant modularity of the CST formulation, including a method for imposing volume constraints.

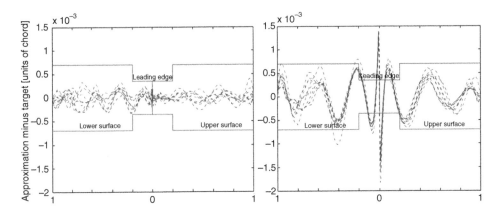

Figure 7.13 CST approximation errors over the set of SC(2) aerofoils using the general aerofoil class function ($\sqrt{x}(1 - x)$) on the right and an SC(2)-family-specific class function on the left. Reproduced from Powell (2012).

7.4 Other Formulations: Past, Present and Future

The aerofoil geometry formulations discussed in this and the preceding chapters were selected to represent a range of philosophies.

In Chapter 3 we introduced the concept of aerofoil geometries based on splines of various flexibilities. They are a sweetshop[2] of parameterizations, on account of the large number of 'controls' they come equipped with and the boundless ways in which they can be expanded by the addition of new control points (though the size of the associated design space tends to have a rather unfortunate response to profligate flexibility enhancement). In this and the previous chapter we also discussed conformal mappings, polynomials and basis functions, as basic principles upon which to build an aerofoil parameterization. Here, we shall briefly mention a small selection of other schemes based on these principles.

Consider, for instance, the following explicit equation defining a parametric 'bump' one may add to a baseline aerofoil:

$$f_b(x) = H[\sin(\pi x^{-\ln 2/\ln x_p})]^t, \quad x \in [0, 1], \tag{7.14}$$

where H is the height of the bump, x_p is the location of the peak of the bump and t is a parameter that controls the width of the bump (larger values of t correspond to sharper bumps). This is a generalized version of the Hicks and Henne (1978) bump functions mentioned in the Preface, and it is capable of altering the shape of a baseline aerofoil in a way that generally maintains its 'aerofoil-like' aspect. In other words, a fairly robust parameterization can be engineered using these bumps, provided that some care is exercised when choosing appropriate ranges for H, x_p and t.

Naturally, if more flexibility is desired, multiple instances of $f_b(x)$ could be added to the baseline, though clearly this makes the robustness control and the optimization process more

---

[2] That is, appealing, but bad for your health if used without caution.

difficult. Once again, the key to effective optimization lies in a good understanding of the balance between flexibility, robustness and conciseness (Section 2.3). Hicks and Henne (1978) derived their bump functions from the shapes of the legacy aerofoils of the time, much like the NACA four-digit aerofoils bear the 'DNA' of the Göttingen 398 and the Clark Y (see the orthogonal basis functions by Robinson and Keane (2001) as a contemporary application of this principle).

The idea of single- or two-piece polynomial descriptions of aerofoil surface shapes has also seen further adoptions beyond the NACA families. Take, for example, the 11-variable model of Sobieczky (1998), which uses a sixth-order polynomial and was proposed for a range of applications, including experiments with transonic aerofoils.

Today, aerofoil design is far from having converged on a universally accepted methodology. Direct and inverse design have their advocates, as do adjoint-based local improvement methodologies and broader scope (more global) techniques based on the parameterization schemes discussed above.

What is fairly certain, however, is that whatever parametric (or perhaps parameterless, 'free'-surface-based) scheme gains most traction in the future, it will have two key features. First, it will offer a means of controlling the flexibility–conciseness–robustness trade-off. Second, it will incorporate some aerodynamics-specific knowledge – a scheme treating an aerofoil like an arbitrary curve without any flow-specific constraints is likely to waste design variables on unnecessary flexibility.

8

Planform Parameterization

Most of us approach the definition, description and analysis of a 3D object via its 2D projections and planar sections. The draughtsman of yesteryear was inescapably committed to this type of approach by the physical constraints of the drawing board, but even today's engineers, in possession of modern 3D CAD tools, will often start by stepping back to 2D – by choice. In fact, most CAD tools encourage a philosophy of building 3D objects via the extrusion of 2D sketches.

Perhaps there will come a time when advances in design technology will have made a direct 3D approach more natural (3D pens and immersive, motion-detection-based CAD tools that support interactions akin to sculpting already point in this direction), but there is more to the role of certain 2D sketches in the aerodynamic design of aircraft than mere drawing expediency. Certain features of some planar sections and projections provide hints as to the performance of the aeroplane that most aerodynamics engineers are able to interpret quite readily. We have so far dealt with one such 2D sketch type: the lifting surface cross-section (aerofoil). This chapter is dedicated to the parameterization of another: the projected *planform* of the wing.

While the planform shape of a wing can be rather complicated (think of curved leading edges or trailing edges made up of multiple straight/curved segments), there are some key numbers that encapsulate the aerodynamic essence of a planform geometry – we shall start with a nondimensional number linked to virtually every aspect of the performance of a wing.

8.1 The Aspect Ratio

Formally, the aspect ratio of a wing is defined as the *span* squared divided by the *projected wing area*, that is, $AR = b^2/S$. The span is the distance between the wingtips, while the projected wing area is the area of the projection of a virtual lifting surface resulting from an imaginary extension of the two wings until they meet at the symmetry plane of the aircraft[1] (more on the wing area in Section 8.4).

---

[1] On some aircraft the two wings do actually form a single aerodynamic component, supported only by a narrow strut; for example, the human-powered aircraft in Chapter 12.

Aircraft Aerodynamic Design: Geometry and Optimization, First Edition. András Sóbester and Alexander I J Forrester.
© 2015 John Wiley & Sons, Ltd. Published 2015 by John Wiley & Sons, Ltd.

Figure 8.1 Aspect ratios from the extremely low – the Vought V-173 'Flying Pancake' (just over one!), with wingtip-mounted propellers – to high – the Alenia C-27J Spartan, bottom left – and low – Alenia Aermacchi M-346 Master (photographs by A. Sóbester).

The term 'aspect ratio' originates from a special case, that of straight, rectangular wings, where AR is what the common usage of the term implies: the ratio of the span and the chord length. The above equation is a generalization to wings of any shape (it can even be applied to oddities like the 'Flying Pancake' seen in Figure 8.1, delta wings, like seen in Figure 8.10 or highly swept wings, like those of the Boeing B-52 shown in Figure 8.2, bottom right), and it preserves the fundamental intuitive meaning of the term: long, slender wings have high aspect ratio and a low aspect ratio indicates a short, stubby wing.

While the above definition seems clear cut, aspect ratio is actually a term rife with ambiguity, which must be eliminated if the aspect ratio is to be used as a geometrical design variable. Both the span and the projected area of many aircraft wings vary throughout the operational cycle. They depend on the quantity of fuel in the wing tanks, as well as on the phase of flight. For example, the wings bend upwards in flight, and it could be argued that aerodynamic calculations should be based on the 'in-flight' projected geometry. The area also depends slightly on the ground stance of the aircraft.

More scope in ambiguity lies in wings fitted with wingtip devices. From an aerodynamics point of view, a winglet is simply a 'folded up' portion of the wing; that is, adding winglets

Figure 8.2 Big jets with high aspect ratio wings. The wings of the Boeing 747-400 (top left, Group V, Code E, span ~55 m) and the Airbus A330 (bottom left, Group V, Code E) are visibly more slender than those of the Airbus A380 (top right, Group VI, Code F, span ~80 m), the latter designed up against the upper boundary of Group VI/Code F. There are fewer geometrical constraints on long-range, strategic bombers, such as the Boeing B-52 (bottom right) (photographs by A. Sóbester).

of semi-span b_{wl} to a wing has an effect comparable to adding $2b_w$ to the span of the straight wing. Should we therefore use the 'span' resulting from adding the winglet semi-spans to the distance between the tips of the 'basic' wing? The answer depends on the purpose of the exercise: for aerodynamics calculations it would make sense, but for calculating airport space requirements (more on which soon), the distance between the tips of the winglets (effectively the largest lateral dimension of the aircraft) would be more germane.

In general, the more unusual the wing configuration, the less clear the definition of AR becomes. Consider, for instance, the leading edge extensions of the F/A-18 (Figure 8.3, left), or a wing with generous fuselage blends: what exactly do we include into the projected area? Logic would dictate that the lift generated by the fuselage should play a part. After all, we are tacitly assuming that the virtual extensions of the wings, 'concealed' by the fuselage, generate the same amount of lift as the exposed sections.

Therefore, it is not surprising that it is often difficult to find reliable (or, indeed, any) figures for the wingspan or projected wing area of existing aircraft, and the uncertainties are obviously even greater for the aspect ratio. While in many applications this is of limited relevance, care must be taken when comparing different aircraft against these metrics or when an optimization process may visit wings with significant differences in topology. If we are optimizing a fixed topology, AR is useful as a rough and ready surrogate for the inverse of induced drag (more

Figure 8.3 Fighter jet planforms: low aspect ratio wings on the Boeing F/A-18E Super Hornet (left) and on the McDonnell Douglas F-15C Eagle (right). Note also the leading edge extensions on the Hornet (photographs by A. Sóbester).

on the reasons shortly) – we just have to stay consistent with whatever definition we choose: we need to stick with it throughout the optimization process.

Aspect ratio is also best regarded, somewhat counter-intuitively, as more of an aerodynamic variable than a geometrical one. Consider, for instance, a pair of high aspect ratio, straight, rectangular wings. Sweeping the wings back by, say, 45° by simply pivoting each wing by that angle will reduce the aspect ratio, as defined above, significantly (the area will stay roughly the same, but the span will decrease), even though the purely geometrical 'aspect ratio' (intuitively seen as the longer side divided by the shorter one) of each wing will stay the same. To the aerodynamicist the change in aspect ratio makes perfect sense, as the span (which is really a measure of the width of the airmass affected by the aircraft) has become shorter with the increase in sweep.

After wing area (to be discussed in Section 8.4), the wing aspect ratio may be the most important design variable in aircraft engineering, in spite of its somewhat fuzzy definition. This is because it is the key driver of one of the fundamental multi-objective trade-offs that define the geometry of any fixed-wing aircraft. Several effects are in competition here; we discuss them in turn next.

8.1.1 Induced Drag

The best known result of the Lanchester–Prandtl lifting line theory is the induced drag equation for planar wings with elliptical lift distributions

$$D_i = \frac{L^2}{\pi q b^2},$$ (8.1)

which tells us that for a given amount of lift L generated at the flow conditions defined by the dynamic pressure q, the induced drag is determined by the span b. From a design perspective, the nondimensionalized version of this equation is more useful, as its terms are more conveniently linked to actual design decisions:

$$C_D^i = \frac{C_L^2}{\pi AR}.$$ (8.2)

Here, C_D^i denotes the induced drag coefficient, with the total drag coefficient being $C_D = C_D^p + C_D^i$, where C_D^p is the parasitic drag coefficient (combining pressure drag, friction drag and wave drag). Aspect ratio is thus intimately related to induced drag (though McLean (2012) warns against losing track of the fact that, from a physical point of view, it is actually the span term of AR that really drives it).

For a given projected wing area, a longer (and thus more slender) wing will produce weaker wingtip vortices than a shorter, wider chord wing. As a result, the lift-induced component of drag (vortex drag) will be greater on short wings.

The Vought V-173 'Flying Pancake' (Figure 8.1) represents a rather unique attempt at reducing the vortex strength associated with a very low aspect ratio: it has wingtip-mounted engines and propellers, rotating in the opposite direction to that of the vortex, thus mitigating its drag-inducing effect. It is also interesting to note that the V-173 had symmetrical four-digit NACA aerofoils (Section 6.1), flying at low speeds and very high angles of attack.

The V-173, however, remains little more than an interesting footnote to the bigger picture of aviation history. Wingtip devices represent a more practical means of dealing with the effects of wingtip vortices for a given aspect ratio – they promise drag reductions of 3–6%. Ultimately, though, such fixes are merely (often problematic, but generally inevitable) compromise solutions – increasing aspect ratio remains the fundamental means of reducing induced drag for a given lift coefficient.

With hardly any wings obeying the assumptions of Equation 8.2 – that is, planar wing with elliptical spanwise lift distribution – a *span efficiency factor e* is needed to account for the corresponding shortfall in lift generation efficiency:

$$C_D^i = \frac{C_L^2}{\pi e \mathrm{AR}}. \tag{8.3}$$

The efficiency penalty of typical deviations translates to an e range of about $[0.85, 0.95]$. Since the other components of the total drag also vary with angle of attack, just like the induced drag does, they can also be regarded as varying with C_L, and they are therefore commonly lumped into the same algebraic form as C_D^i above, leading to a total drag expression of the form

$$C_D = C_{D0} + \frac{C_L^2}{\pi e_0 \mathrm{AR}}, \tag{8.4}$$

where C_{D0} is the zero-lift drag (simply the overall drag at the angle of attack at which no lift is generated) and e_0 is the *Oswald efficiency factor*. This is not to be confused with the e of Equation 8.3, as it refers to all α-dependent components of the overall drag. e_0 is usually in the range $[0.7, 0.85]$ (considerably lower at supersonic speeds). Here is a set of empirically derived equations, dating back to the 1960s and commonly used ever since (e.g. see Raymer (2006)), for an initial estimate of e_0:

$$e_0 = \begin{cases} 1.78(1 - 0.045\mathrm{AR}^{0.68}) - 0.64, & \Lambda_{\mathrm{LE}} \leq 30° \\ 4.61(1 - 0.045\mathrm{AR}^{0.68})(\cos \Lambda_{\mathrm{LE}})^{0.15} - 3.1, & \Lambda_{\mathrm{LE}} > 30° \end{cases}, \tag{8.5}$$

where Λ_{LE} denotes the leading edge sweep.

8.1.2 Structural Efficiency

Low aspect ratio wings are structurally more efficient. Most wings are cantilevers that carry an approximately elliptically distributed load, which adds up to the product of the mass of the aircraft and the g-loading. The longer the cantilever carrying this load is (i.e. the greater the aspect ratio), the heavier the required spars will be, which leads to an increase in the overall weight of the aircraft. This, in turn, will increase the amount of lift needing to be generated, which, in turn, will increase the required wing area... Even if the design process does not quite spiral out of control in this manner, it is possible that some of the induced drag gains will be cancelled out.

8.1.3 Airport Compatibility

Airport compatibility constraints often dictate a maximum limit on the span, and thus, for a wing area calculated performance constraints (more on which in Section 8.4.1), on the aspect ratio. In order to simplify the process of determining the number of airports an aircraft of given dimensions would be able to operate from, a set of six design groups have been drawn up by the International Civil Aviation Oragnization (ICAO Annex 14, Aerodome Reference Code Element 2, Table 1-1) and by the Federal Aviation Administration (FAA Advisory Circular 150/5300-13). A set of requirements related to taxiway spacings, runway widths, taxiway–runway separations, and so on correspond to each of the six groups, and so do a set of relevant aircraft design variables too.

Table 8.1 lists the wing span, tail height and outer main gear wheel span ranges corresponding to each code/group. This tells us, that, for example, in order to be able to operate from a Group III-capable airport, the wing span of an aircraft must not exceed 36 m, the tail must not be taller than 13.7 m and the span of its outer main gear wheels should be less than 9 m. The medium-haul workhorses of the airline industry, the Boeing 737 family and the Airbus A320 family, are typical examples of FAA Group III airliners.

A particularly noteworthy number in Table 8.1 is the upper limit on the span of Code VI/Group F aircraft: 80 m (airports in this category are also expected to have runways at least

Table 8.1 ICAO Aeroplane Design Codes/FAA Airplane Design Groups and corresponding geometrical constraints.

ICAO code	FAA group	Span (m)	Tail height (m)	Main gear span (m)	Example aircraft
A	I	<15	<6.1	<4.5	Lear 45, Cessna Mustang
B	II	15–24	6.1–9.1	4.5–6	Saab 340, CRJ-200, ERJ-145
C	III	24–36	9.1–13.7	6–9	A320, B737, ATR-42, Dash-8, MD90
D	IV	36–52	13.7–18.3	9–14	MD-11, B757, B767, A300, A310
E	V	52–65	18.3–20.1	9–14	B747-400, B777, B787, A330, A340
F	VI	65–**80**	20.1–24.4	14–16	A380, B747-8

Note: We have aligned ICAO Aeroplane Design Codes (first column) and FAA Airplane Design Groups (second column) here, as they are basically identical, with only extremely rare examples of aircraft classified differently by the two coding schemes.

60 m wide and separated from the nearest taxiway by at least 190 m; the separation between taxiways must be at least 97.5 m). This effectively means that a practically useful airliner must have a span less than 80 m, and the first (and to date, only) airliner to have come up against this limit is the Airbus A380. The almost exactly 80 m span of the aircraft indicates that the design of its wing lies up against the airport compatibility constraint, and it is likely that a higher than desired aspect ratio is one effect of hitting this boundary. We have discussed the limitations of making comparisons on the basis of actual aspect ratio values, but the AR of the A380 is generally quoted around 7.5, and this is significantly lower than that of comparable airliners, such as the Boeing 747-400 (values closer to 8; see a visual comparison in Figure 8.2) or the Airbus A340-500/-600 (around 9).

The sizes of elevators and ship decks on aircraft carriers impose, of course, far more stringent span constraints on naval aircraft, so these, in spite of having generally much lower aspect ratios anyway (for structural and handling reasons), have to resort to the expensive and heavy solution of folding wings. Interestingly, the drive towards extremely low fuel burn, and thus very low induced drag, might impose similar technologies on commercial airliners too (e.g. see the folding wings of Boeing's SUGAR Volt).

8.1.4   Handling

Tip Stall

Consider an aircraft flying a steady, coordinated turn, at an airspeed (measured near the centre of gravity) of V_{CG}. If the span of the aircraft is b, the bank angle is ϕ and the radius of the turn is R, the airspeeds measured at the inside and outside wingtips (denoted by V_{in} and V_{out} respectively), can be calculated as

$$V_{in} = V_{CG} \left(1 - \frac{b}{2R} \cos \phi \right)$$
(8.6)

and

$$V_{out} = V_{CG} \left(1 + \frac{b}{2R} \cos \phi \right).$$
(8.7)

To gain a feel for what this means in practical terms, consider a glider with a 20 m span in a 45°, 75 m radius turn. If the airspeed indicator, using a signal from a fuselage-mounted Pitot-static system, reads 50 KIAS, from Equations 8.6 and 8.7, the airspeeds at the inside and outside wingtips will be around 45 KIAS and 55 KIAS respectively. This is an illustration of how aircraft with high aspect ratio wings and low wing loadings can become vulnerable to *tip stall* on the slower wingtip, often followed by a *spin*.

For a more detailed treatment the reader may wish to consult Phillips (2010), who notes that at greater aspect ratios this spanwise variation of airspeeds may also be significant from the point of view of the *stall-limited maximum load factor* (i.e. the maximum number of g achievable in a turn on the stall boundary). This is usually calculated as

$$n_{max} = \frac{q}{W/S} C_{L_{max}},$$
(8.8)

where q is the dynamic pressure measured by a centrally mounted Pitot-static system, W/S is the wing loading and $C_{L_{max}}$ is the maximum lift coefficient. The critical n_{max} on a high aspect ratio wing will be determined by the speed of the flow over the wingtip on the inside of the turn, so

$$n_{max} = \frac{q}{W/S} \left(1 - \frac{b}{2R} \cos \phi_{max}\right)^2 C_{L_{max}}. \tag{8.9}$$

Adverse Yaw

A typical turn is initiated by deflecting the aileron on the wing on the outside of the turn downwards and the inside aileron upwards. This will increase lift on the outside wing and reduce lift on the inside, thus producing the desired roll. There are, however, some unwanted side effects too.

The by-product of the increased outboard lift will be increased induced drag, accompanied by a reduction in induced drag on the inside. At the same time, as the outboard wing starts rising, its angle of attack will change slightly, tilting the lift vector backwards. Both effects will generate moments tending to yaw the aircraft *out of the turn*. The greater the span, the more pronounced this *adverse yaw* effect will be, exacerbated by the fact that ailerons tend to be positioned far outboard in order to maximize their authority – this will, however, increase their 'adverse yaw authority' too.

Phillips (2010) notes that on aeroplanes with very large aspect ratio wings the spanwise airspeed variation caused by the adverse yaw can be so pronounced that the resulting loss of lift on the rising (outside) wing can exceed the increase in lift generated by the downward deflection of the outside aileron, thus initiating a turn in the opposite direction to that normally expected.

A number of fixes are available, especially on fly-by-wire aircraft, where a rudder deflection can be 'mixed in' to cancel out the adverse yaw. Other solutions include the asymmetrical deflection of the ailerons (i.e. greater deflection upward than downward) and *Frise ailerons*, which pivot in such a way that their leading edge dips under the wing when deflected upwards. All such solutions, however, have their drawbacks and do not change the fundamental fact that adverse yaw increases with increasing span, placing yet another limit on this very effective induced drag reduction measure.

8.2 The Taper Ratio

The taper ratio is simply defined as $\lambda = c_{tip}/c_{root}$; that is, the tip chord length divided by the root chord length. λ is usually less than one (see the right image in Figure 8.4 for a very rare exception[2]), a reflection of the main reason why wings tend to be tapered: it is a simple way of increasing the span efficiency factor (tapering enables a closer approximation of an elliptical lift distribution). A secondary advantage is that it can improve the structural efficiency of the wing by moving more of the load close to the root.

---

[2] Raymer (2006) cites a possible reduction in wing–fuselage interference drag and avoidance of tip stall as the apparent reasons for the unusual choice on the XF-91.

Figure 8.4 Tapered wings. Sub-unity taper ratio on the Shorts Tucano (left, $\lambda < 1$) and reverse taper ($\lambda > 1$) on the wings of the Republic XF-91 Thunderceptor (photographs by A. Sóbester and US Air Force).

8.3 Sweep

8.3.1 Terminology

Most aeronautical engineers have little difficulty identifying a *swept* aircraft wing – that is, one with a nonzero wing *sweep angle* – but the latter term is dangerously ambiguous, and it is used often as a lazy shorthand for a number of different geometrical quantities.

The *leading edge sweep angle* Λ of a wing with a straight leading edge is the angle between the planform projection of the leading edge and a line perpendicular to the axis of symmetry of the planform projection. The concept of a 'straight leading edge' can be stretched somewhat for the purposes of this definition. For example, root or tip fillets can be ignored if the leading edge is otherwise straight.

If the leading edge is made up of two segments with a nonzero angle between them, the wing can be characterized by an inboard and an outboard sweep angle. For instance, the main wing leading edge of the Grumman X-29 (Figure 8.5) comprises a backwards swept and a reverse swept segment; that is, the inboard leading edge sweep angle is positive and the outboard leading edge sweep angle is negative.

Let us also introduce here the concept of the *local leading edge sweep angle* $\Lambda(y)$ or $\Lambda(\varepsilon)$. This is the angle between the tangent to the wing planform leading edge and a lateral axis perpendicular to the axis of symmetry of the planform projection. The spanwise variation of this angle (as a function of the Cartesian coordinate y or a leading edge-bound coordinate ε, more on which in Section 9.2.2) defines the shape of the leading edge, and it is a concept that becomes necessary when defining wings with curved leading edges (see Figures 8.1 and 9.6 for examples).

Similar terminology can be defined at chord stations other than zero (i.e. other than the leading edge). For instance, replacing the leading edge with the geometrical locus of the quarter-chord points of the aerofoil sections that make up the wing, one could define a *quarter-chord sweep angle* (denoted $\Lambda_{1/4}$ or $\Lambda_{0.25}$) or a *local quarter-chord sweep angle*.

Figure 8.5 Forward-swept outboard wing segments on the experimental Grumman X-29 (photograph by US Air Force).

Another pair of terms used in association with swept wing is *apparent span* or *structural span* b_s. This is simply the length of the leading edge: $b_s = b/\cos(\Lambda)$. The latter name alludes to the way in which engineers typically think of designing swept-wing aircraft; that is, starting with a straight wing and pivoting it backwards or forwards with the internal structure unchanged. The former term is of importance to pilots and airport planners, and it refers to a straight leading edge swept wing appearing, from a turning clearance point of view, to have a span equal to its leading edge length.

Swept wings are typically associated with high-speed (transonic or supersonic) flight (see Figure 8.6 for an early example), though high-speed aircraft do sometimes have zero-sweep wings (see Figure 8.7) and many low-speed aircraft have swept wings. The existence of the latter category is mainly the result of stability considerations, in particular in the case of 'flying

Figure 8.6 The Boeing B-47A, the world's first swept-wing bomber (photograph by US Air Force).

Figure 8.7 SNCASO Trident I. This 1950s experimental aircraft powered by turbojets and a rocket was able to achieve Mach 1.55 in spite of having zero sweep (photograph by A. Sóbester).

wing'-type aircraft, which, in practice, can only be made statically stable via nonzero sweep. Their directional stability might also demand that their wingtips are further aft – this can also be achieved by increased sweep.

For a given wing area, increasing sweep will decrease span (and aspect ratio). As we have seen in Section 8.1.1, this has the effect of increasing the induced drag. Yet, sweep has the potential to reduce drag significantly, especially at higher speeds, where the induced drag coefficient diminishes and wave drag becomes the more important component. More specifically, on transonic aircraft, sweep can be seen as a means of delaying the onset of compressibility effects – we discuss this in more detail next.

8.3.2   Sweep in Transonic Flight

As the speed of an aircraft increases, there comes a point when regions of supersonic flow begin to appear around the wings. The critical free-stream Mach number M_∞^{crit} that marks the occurrence of this *transonic* regime is typically in the range $[0.7, 0.8]$. Compressibility effects begin to appear here, the wave drag component starts to become significant and, very soon after, at the *drag divergence Mach number* M_∞^{dd}, the overall drag begins to rise sharply with Mach number (the alternative term *drag rise Mach number* is sometimes used – we have already touched upon these concepts in the context of supercritical aerofoils in Section 5.2.3). The dissipative nature of the suddenly appearing shockwaves and the separation caused by their sharp adverse pressure gradient are the causes of this phenomenon (Anderson, 2005).

One of the great aerodynamic advances of the 1940s was the observation that sweeping a wing was a simple, practical way of delaying this drag rise. More specifically, it was found that the flow field over a swept wing was determined by the Mach number component that is normal to the leading edge, as well as by the angle of attack in a plane normal to the leading edge. This makes intuitive sense, as only the normal component encounters a significant wing camber and/or thickness increase – on most swept wings the spanwise component will merely 'see' a steady reduction of thickness towards the wingtip.

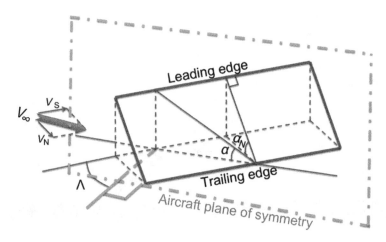

Figure 8.8 Geometry of a swept wing: a normal and a streamwise section through the plane of a straight, plane wing pivoted to obtain a sweep Λ.

Figure 8.8 shows this resolution of the free-stream flow velocity vector V_∞ (corresponding Mach number M_∞) into a spanwise component V_S and a normal component V_N (corresponding Mach number M_N) on a straight, plane wing rotated around a vertical axis by a sweep angle Λ. It can be seen that the angle of attack α_N in a plane normal to the leading edge is greater than the angle of attack α in a streamwise plane and it can be calculated as

$$\alpha_N = \arctan \frac{\tan \alpha}{\cos \Lambda}. \qquad (8.10)$$

Similarly, the normal component of the free-stream Mach number can be calculated as

$$M_N = M_\infty \cos \Lambda, \qquad (8.11)$$

which implies that sweeping the wing should also increase the critical Mach number by a factor of $\cos^{-1} \Lambda$.

All of the above, however, is based on a purely geometrical reasoning, assuming, essentially, an infinite wing. However, the aerodynamics of the 3D flow over a finite swept wing – in particular, a swept wing attached to a body – is considerably more complex than Equations 8.10 and 8.11 would suggest. The wing tends to 'unsweep' the isobars near the root and the tip, and shocks and supersonic regions may be induced by the wing, contributing to significantly greater Mach numbers than those predicted by our earlier naive trigonometry.

Mentioning these pitfalls, Kulfan (2007) also notes that the mid-span region of the wing does tend to be modelled relatively well by the 'infinite wing' theory, and this is confirmed by the fact that the drag response of a swept wing still improves considerably with increasing Λ in the high transonic region. She cites the example of the zero lift drag of a series of similar wings with increasing angles of leading edge sweep, where the drag divergence Mach number rises significantly with Λ, as does the maximum drag Mach number, while the maximum drag coefficient falls – Table 8.2 summarizes this data set.

Table 8.2 Zero lift drag characteristics of similar wings at varying angles of leading edge sweep Λ (data based on Kulfan (2007)).

Root	Tip	AR	Λ (deg)	M_∞^{dd}	$C_{Dmax}^{C_L=0}$	$M_{C_{Dmax}}$
65A010	65A010	6.4	10	0.82	0.076	1.01
65A011	65A008	3.7	31	0.88	0.059	1.02
65A011	65A008	5.1	40	0.92	0.042	1.03
65A010	65A008	5.1	45	0.94	0.038	1.04
65A008	65A008	3.0	49	0.97	0.028	1.05

8.3.3 Sweep in Supersonic Flight

The disturbance caused by a flying object will propagate at the speed of sound. If the object itself travels faster than sound, the wave fronts generated by it will form a cone, beyond which information will not travel from the object. In other words, the free stream outside of the cone will not be affected by the approach of the object. The half-angle of the cone is exclusively a function of Mach number ($\mu = \arcsin(1/M)$) and is referred to as the *Mach angle*. Figure 8.9 illustrates the concept. While sweep is advantageous in the case of supersonic flight too (just as it was in the transonic regime), its effect is subtly different here. More specifically, the wing of an aircraft flying in a supersonic free stream will 'see' a *sub*sonic

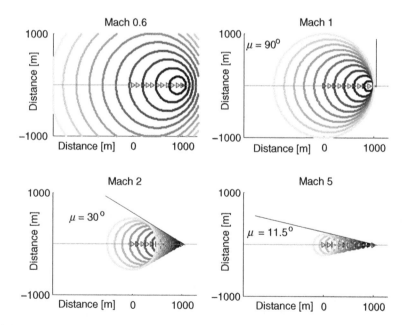

Figure 8.9 Consider an aircraft flying horizontally at sea-level conditions and different Mach numbers, from the zero abscissa to the 1000 m abscissa point. The small triangles spaced at 100 m represent its path and the circle centred on each triangle is the wave front (recorded at the moment the aircraft reached the end of the 1 km segment) of the disturbance created by the aircraft in that point. The aircraft has no effect on the free stream outside of the cone defined by the wavefronts.

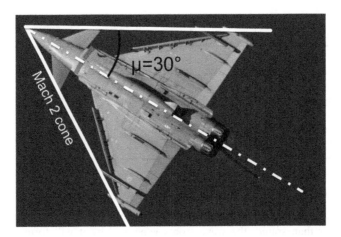

Figure 8.10 Eurofigher Typhoon – the Mach 2 cone of the nose superimposed upon the planform of the aircraft ($\mu = \arcsin(1/2) = 30°$).

normal component *if the leading edge is inside the Mach cone.* This condition is often the key design driver that determines the sweep angles of the wings of supersonic aircraft (see Figure 8.10).

8.3.4   Forward Sweep

Throughout the history of fixed-wing aviation a number of designs teams were driven to small absolute values of negative Λ by relatively mundane desires, such as to correct slight problems with stability, handling or main spar positioning. The more interesting question is: is there ever a good reason for more deliberate and more substantial forward sweeping of a main wing?

The fundamental idea of reducing the normal component of the flow velocity by sweeping the wing applies equally to wings swept back and forward. However, most swept wing, high–speed aircraft feature backward sweep ($\Lambda > 0$), mainly due to the following *disadvantages of forward sweep*:

- *Divergent aero-elastic behaviour* – the aerodynamic loads on the outboard section of the wing will twist it in a way that increases the angle of attack near the tip. In turn, this will further increase the aerodynamic loads that caused the twist in the first place. An extremely stiff wing structure is therefore needed, which increases cost and weight.
- *Unstable stall behaviour* – when the root of a forward swept wing stalls first, a pitch up will result. This will increase the angle of attack, worsening the stall. In contrast, aircraft with wings with no sweep or positive sweep tend to drop their noses following a stall, which will increase the speed and decrease the angle of attack, thus facilitating recovery.
- *Yaw instability* – this is yet another divergent behaviour; yawing reduces the apparent sweep of the wing on the side of the direction of the yawing motion, thus increasing its drag at high speeds. This has the tendency to further increase the yaw angle. The opposite is true of backward swept wings, where yaw causes a restoring moment.

- *Forebody Mach cone* – the higher the design Mach number is, the smaller the angle of the Mach cone. If we are to keep the leading edge within the cone, the span-limiting effect of this constraint is greater than in the case of positive sweep.

Nevertheless, some key *advantages of forward sweep* may make it an attractive proposition in some cases:

- *Layout* – a key problem on smaller pressurized aircraft is the location of the main spar with respect to the pressure cabin, which it cannot go through for structural reasons. The most common solution is to pass the wing underneath (or, less frequently, above) the cabin, but this increases fuselage cross-sectional area and requires complicated wing-to-body fairings. The alternative is a mid-wing passing aft of the pressure cabin, but this may lengthen the fuselage excessively (pushing the centre of pressure far aft) – unless the wings are swept forward! The HFB 320 Hansa Jet, a 10-seat business jet developed by Hamburger Flugzeugbau in the 1960s, features this solution. Similarly, forward sweep can be used to increase the unobstructed fuselage space available near the centre of gravity. This piece of real estate is especially important in a bomber, the centre of gravity of which should not change as bombs are released. This was the design thinking behind Nazi Germany's Junkers Ju-287 forward swept-wing experimental bomber.[3]
- *The inboard flow* over the wing is the key aerodynamic advantage of the forward swept wing. This becomes especially significant at low speeds and high angles of attack, when separation may have already led to the stall of the inboard section when the flow is still attached near the wingtips and, most importantly, over the ailerons. Such high angle of attack capabilities were the main design driver behind the forward swept wing of the X-29 (Figure 8.5). This turned out to be a very successful research aircraft, having demonstrated a very broad angle of attack envelope (see Figure 8.11). While some of the boundaries of this envelope have probably been pushed out by very favourable interactions with the flow over the forebody and the canard (Johnsen, 2013) and the handling characteristics in the regions beyond 40° were found to be somewhat challenging (especially in terms of yaw control), the X-29 confirmed the high angle of attack potential of the forward swept wing.
- *Reduced twist*. Aft swept wings typically feature *washout* (downward twist at the tip – more on twist in Section 9.1.1) to delay tip stall after the root has stalled. While this maintains controllability at low speeds (keeping the ailerons unstalled), it increases cruise drag at high speeds. Owing to the opposite direction of the spanwise flow component on forward swept wings, these do not require such twist (or, sometimes, a small amount of upward twist), which reduces drag at high speeds (Johnsen, 2013).

8.3.5 Variable Sweep

We mentioned in the introduction to this section that sweeping a wing (while maintaining its area constant) will reduce the span and the aspect ratio and this has a detrimental effect on induced drag. Aircraft with significantly swept wings will therefore pay a penalty for their

---

[3] The chief engineer of the Ju-287 project, Hans Wocke, would go on to lead the HFB 320 Hansa Jet design team too, nearly two decades later (Johnsen, 2013).

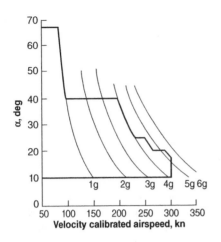

Figure 8.11   The angle of attack envelope of the X-29 forward swept-wing research aircraft (angle of attack versus airspeed at a range of load factors) and the aircraft in high angle of attack flight (as shown by the smoke generated for visualization purposes) – both courtesy of NASA, plot from Bauer et al. (1995).

good high-speed performance in the lower reaches of their speed range, where the induced component dominates the total drag. This is a particularly significant issue when the mission profile of an aircraft combines periods of low-speed loiter with high-speed dash segments – strike aircraft are a typical example (we have discussed the challenges and techniques of such multipoint design in Section 2.7.2).

This reasoning suggests that there is an optimum sweep angle for every speed (at least beyond a certain threshold speed or Mach number value). Therefore, if an aircraft has to spend comparable amounts of time in low-speed, high-endurance loiter and high-speed dash and/or relatively fast cruise, an argument could be made for a variable-geometry wing – specifically, a wing whose sweep varies according to some schedule linked to Mach number. If a continuous sweep angle adjustment is mechanically possible, the optimal schedule is likely to be similar to an arccos law derived from Equation 8.12; that is

$$\Lambda(M_\infty) = \arccos \frac{M_N^t}{M_\infty}, \tag{8.12}$$

where M_N^t is some target normal section Mach number we are aiming to maintain throughout the speed range of the aircraft.

The cost, complexity, structural weight, reliability and maintainability penalties of variable sweep tend to be high, but, depending on the exact shape of the mission profile, it might pay its way in terms of aerodynamic efficiency, and thus mission performance. Such reasoning led to the design of the General Dynamics F-111 Aardvark (the first production aircraft to feature a variable-sweep wing), the Grumman F-14 Tomcat, the mighty Rockwell B-1 Lancer supersonic bomber and the Panavia Tornado (Figure 8.12). A concept referred to as the *oblique wing* (or *yawed wing*) is a clever twist (as it were) on the variable-sweep idea. It involves a single wing, with a continuous spar (circumventing some of the most serious structural penalties of other variable-sweep concepts), which can pivot to a wide range of angles of sweep – forward

Figure 8.12 Panavia Tornado ADV (F3), a Mach 2.2 long-range variable-sweep interceptor with its wings at $\Lambda = 67°$, their highest sweep setting, the lowest being $\Lambda = 25°$ (photograph by A. Sóbester).

sweep on one side and backward sweep on the other (see Figure 8.13). In addition to the structural benefits, the oblique wing also maintains the location of the centre of pressure roughly constant, thus eliminating one of the major difficulties with symmetrical designs.

However, there are drawbacks too (which explains the absence of major oblique wing projects since the late 1970s). First, such designs may experience some of the problems of forward swept-wing aircraft (such as aero-elastic divergence, as outlined in Section 8.3.4, though Kulfan (1973) did not find these to be a limiting factor in a major NASA/Boeing oblique wing airliner study in the 1970s). Additionally, engine integration and landing gear design present greater difficulties than on more conventional aircraft.

8.3.6 Swept-Wing 'Growth'

Pilots of swept-wing aircraft are aware (some painfully so) of a phenomenon known as *swept-wing 'growth'*. This is the apparent increase of the span in terms of turning clearances on taxiways, experienced on swept-wing aircraft with wingtips positioned aft of the main landing gear. Figure 8.14 is a sketch of the arc drawn by the wingtip of a turning swept-wing aircraft in the extreme case of the port side main landing gear being braked to minimize the turn radius and, therefore, becoming the pivot point.

Figure 8.13 NASA oblique wing aircraft: OWRA (Oblique Wing Research Aircraft) unmanned demonstrator (left) and the AD-1, a subsonic, manned research aircraft from 1979 (NASA images).

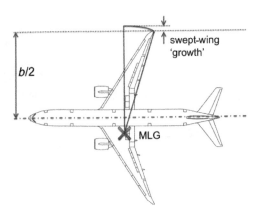

Figure 8.14   Swept-wing aircraft turning on a taxiway around its port main landing gear (MLG). The space required is greater than the span of the aircraft, by a margin known as *swept-wing 'growth'*.

The starboard wingtip can be seen to exceed the track defined by the span of the aircraft. This is of importance when travelling on a curved path through the space between two obstacles only slightly more than a span apart (e.g. entering a hangar).

8.4   Wing Area

It is hard to overstate the importance of the choice of wing area as part of the conceptual and preliminary design of aircraft. For a start, in the spirit of the discussion in Section 2.1, this is the variable that, for the first time in the design process, *scales* a parametric geometry.

It also has a unique distinction: while it is not an objective function per se, it is such a good surrogate of several objectives – most notably wing weight and cost – that it is often actually regarded as one. Many classic design heuristics are centred around the idea of minimizing wing area.

Of course, given an optimization framework equipped with a complete set of dependable analysis capabilities, we would set a higher level objective function (such as life cycle cost or perhaps fuel burn per passenger mile – recall the discussion on the hierarchy of objective functions in Section 2.7.1) and we would let the optimizer select all variables, including wing area. Such multidisciplinary design optimization capability, however, may be a luxury in many design scenarios, in particular in the early phases of the design process, in which case minimizing the wing area itself is a sensible option.

In either case, however, we need to identify the constraint boundaries, which delimit the design space of a wing, and we shall discuss these next.

8.4.1   Constraints on the Wing Area

Wing Loading

The role of the wing is to generate the lift required to oppose the weight of the aircraft. Larger wings tend to generate proportionally more lift (hence the first-order area term in the lift

Table 8.3   Typical wing loading value ranges.

Aircraft category	Wing loading range (N/m^2)	Wing loading range (lbs/ft^2)
Light aircraft	50–150	10–30
Commuter props	100–300	20–60
Business jets	150–500	30–100
Jet airliners	500–900	100–180
Fighter jets	300–700	60–140

Note: the values listed here refer to the maximum take-off weight of each class of aeroplane.

equation!), so it is a fairly obvious observation that the wing size scales with weight, *other things* being equal. Thus, it makes sense to combine area and weight into a single, pressure-like term, which neatly gets rid of this scale effect and lets us focus on those *other things*: the effect of wing area on the maximum rate of turn, rate of climb, gust response, and so on. This term is the *wing loading*, defined simply as W/S, the weight at the flight condition being considered, divided by the projected area of the wing.[4] It is typically measured in newtons per square metre, or, more rarely, pounds per square foot or kilograms per square metre (the latter is, of course, a slightly different type of quantity: mass per unit wing area). Table 8.3 shows representative value ranges for a few common classes of aircraft.

Thrust-to-Weight and Power-to-Weight Ratios

In the interest of conciseness, we shall, in the following, discuss constraint analysis in the context of thrust-to-weight ratios (versus wing loading). This obviously works for jet engines, but we can use it (with extra care) for propeller-engined aircraft too. We need to note that the thrust T generated by propeller engines depends on the density altitude (usually characterized for performance calculation purposes by the density ratio ρ/ρ_0; that is, the ambient air density divided by the sea-level density), as well as on the airspeed V, and is usually estimated as

$$T = P \frac{\rho}{\rho_0} \eta \frac{1}{V}, \qquad (8.13)$$

where P denotes the maximum power available. Further caution is advised in terms of understanding the power versus altitude curve of the engines, in particular when they are flat rated; that is, when an artificial limit is placed on the peak power output of the engine under certain environmental conditions. η is the efficiency of the propeller (typical values around 0.8).

---

[4] Technically, this should be the projection onto a plane parallel with the free-stream lines, though correction for variations in angle of attack generally makes little difference.

Constraint Analysis

Perhaps the most interesting interplay between thrust-to-weight ratio and wing loading occurs in rate/angle-of-ascent-related constraints. As shown by Raymer (2006), the relevant constraint boundary is of the form

$$\frac{W}{S} = \frac{T/W - G \pm \sqrt{(T/W - G)^2 - 4C_{D_0}/\pi ARe}}{2/q\pi ARe} \tag{8.14}$$

subject to

$$\frac{T}{W} \geq G + 2\sqrt{\frac{C_{D_0}}{\pi ARe}}. \tag{8.15}$$

The latter condition ensures real-valued solutions; as Raymer (2006) points out, the physical significance of this inequality is that, no matter how aerodynamically efficient the aircraft is expected to be, the thrust-to-weight ratio must exceed the climb gradient. In the above equations, W/S and T/W denote the wing loading and the thrust-to-weight ratio respectively, G is the climb gradient (slope), C_{D_0} is the zero-lift drag, AR is the wing aspect ratio and e is the efficiency factor.

Certification criteria and design briefs usually include a number of climb-angle- or rate-of-climb-related constraints. Consider, for example, the *initial climb* requirement. This typically imposes an angle of climb on an aircraft ascending from runway level with landing gear down, sometimes following the failure of the critical engine. Figure 8.15 features a grey and a white region, separated by a boundary computed using Equation 8.14. Thrust-to-weight ratio and wing loading pairings in the grey region define aircraft that would fail to make the required climb slope target.

Such constraint diagrams are also useful sensitivity analysis tools. For instance, if we could relax the climb angle requirement by 1°, how would that change the location of the boundary? What about the effect of a more severe climb requirement, exceeding the nominal value by 1°? The medium and heavy continuous black lines in Figure 8.15 depict the answer. Similarly, what would be the impact of changing the target speed? In the case of the initial climb, flight manuals typically refer to this value as V_2; the heavy and medium dashed lines show the constraint boundaries corresponding to increasing or decreasing V_2 by 15 knots.

Equation 8.14 can also be used to compute the *service ceiling* requirement constraint boundary. This is usually expressed as a rate of climb (typically of the order of 100 m/min) to be attainable at the density altitude that is to be designated as the aircraft's service ceiling.

Another typical thrust-to-weight ratio–wing area constraint family is that related to take-off performance. Take-off roll requirements, a specified distance up to the nearest point where an obstacle of a certain height can be cleared on take-off, maximum balanced field length and variations thereof can typically be expressed as constraints of the type shown in Figure 8.16.

We shall see examples of these constraints in the case of a 'real-life' wing design problem in Section 9.4. For now, let us consider Figure 8.17, a scatter plot positioning a number of aeroplanes in thrust-to-weight versus wing loading space. More specifically, the location

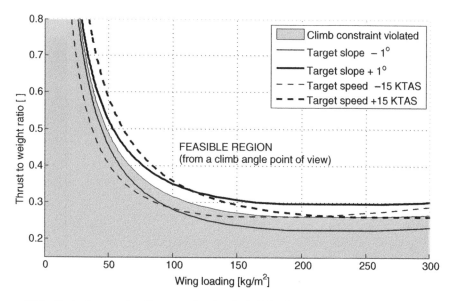

Figure 8.15 Typical constraint diagram reflecting the initial climb requirement. The curve separating the feasible (white) and the infeasible (grey) region of the thrust-to-weight versus wing loading space is the nominal constraint boundary. The additional boundaries show the impact of changes to the nominal (target) values of some key parameters on the location of this boundary.

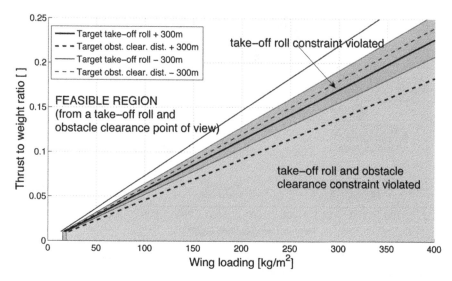

Figure 8.16 Typical constraint diagram reflecting take-off distance requirements. The larger of the two superimposed grey triangles is the zone in which a design would violate the constraint on the maximum take-off roll length. The continuous lines either side of the corresponding boundary indicate the sensitivity of the location of this boundary to the target length. The 50 ft obstacle clearance feasibility constraint (which can be read off the graph in a similar way) is less restrictive, and therefore inactive in this particular case.

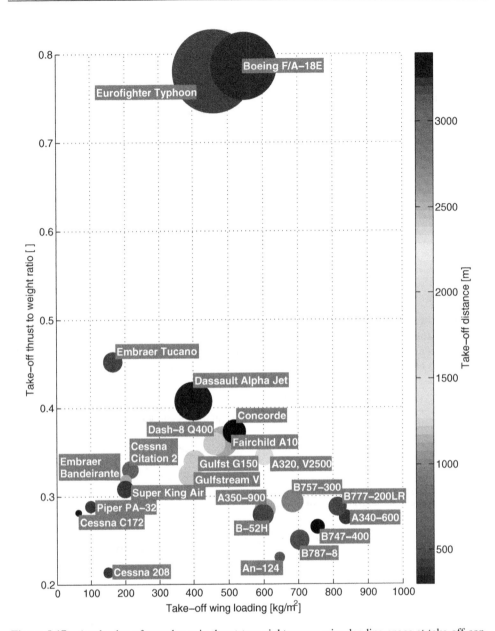

Figure 8.17 A selection of aeroplanes in thrust-to-weight versus wing loading space at take-off conditions. The colours of the discs representing each aeroplane indicate the length of their International Standard Atmosphere (ISA), maximum take-off weight and take-off distance, while their size is proportional to their initial rate of climb under the same conditions.

of these points was calculated on the basis of take-off weight and take-off thrust. A number of propeller-driven aircraft are shown too – the thrust-to-weight ratio of these was calculated on the assumption that their engines produce their maximum continuous sea-level power at take-off speed. In all cases, ISA sea-level conditions were assumed.

This is not a particularly precise comparison – we have seen that even something as straightforward as the wing area may be reported slightly differently by different manufacturers, and there are numerous similar factors involved here (to mention just one more, some of these aircraft are available with different types of engines – in these cases we used an average thrust value). It is nevertheless indicative of which categories of aircraft populate various regions of the thrust-to-weight ratio versus wing loading space.

In addition to the distribution of this range of designs, the colouring of the discs representing each aircraft – proportional to their take-off distances – and the sizes of the discs – proportional to the initial peak rates of climb – reflect some of the features we saw on the constraint plots in Figures 8.15 and 8.16. Again, the performance metrics plotted here should be considered as approximate numbers, for the reasons indicated above, but the general trends are still clear.

8.5   Planform Definition

8.5.1   From Sketch to Geometry

Early on in this book, in Section 2.1, we saw how the physics of airflow around objects encourages the separation of shape and scale when it comes to understanding the key features of the flow field. Applying the same reasoning to the design process is by no means mandatory, but it often makes sense.

Engineers do this naturally, especially in the early stages of design processes. The legendary 'back of the napkin' drawings of famous designs – think Roy Chadwick and the Avro Vulcan or Burt Rutan and the Voyager – rarely feature dimensions; they capture the essence of the design in a nondimensional sketch. Chadwick had a reasonably clear idea of the shape of the leading edge of the wing of the Vulcan – that is, the spanwise variation of the local sweep angle $\Lambda(\varepsilon)$ (where ε is some spanwise coordinate) – long before he could assign a precise figure to its span or area.

Designing in terms of nondimensional shapes also makes designing aircraft *families* more natural – a nondimensional baseline design can be instantiated to various scales determined by the individual performance requirements of the members of the family (this is a scenario typical of small, low-cost unmanned air vehicles – the scaling of families of larger, more expensive aircraft is heavily influenced by commonality constraints, which often preclude geometrical similarity; for instance, members of the same family of airliners will often share the same wing and fuselage frames, but will have different fuselage lengths).

Finally, we may also benefit from this scaling approach when it comes to engineering the computer code that will generate the geometry. Consider, for example, a class called `Lifting-Surface`, which will have a number of inputs, such as designer-supplied methods defining the shape of the leading edge (basically $\Lambda(\varepsilon)$), the variation of chord length with spanwise station, and so on. The class could also include a method called `GenerateLiftingSurface`, which takes a scaling factor as its argument, and it does what it says on the tin: actually builds the geometry at the given scale. We can then instantiate the class with a given nondimensional shape definition, and once this is done we can call `GenerateLiftingSurface` as many times as we like at different scales, as dictated by an optimizer, a design study or our whim. Indeed, this is the design philosophy (complete with method naming) we applied to the OpenNURBS/Rhino-Python-based `LiftingSurface` code accompanying this book. We shall introduce this in detail shortly, but first we need to look at some of the details of scaling a wing geometry.

8.5.2 Introducing Scaling Factors: A Design Heuristic and a Simple Example

From an engineering point of view, the most convenient aspect of scale is the wing area, as a result of its intimate connection with almost every aspect of performance. In Section 8.4.1 we reviewed the principles of thrust-to-weight ratio versus wing loading constraint analysis and we saw how this could yield the appropriate wing area for the aeroplane being designed (in Chapter 9 we will illustrate this via a numerical example). The initial geometry could thus result from the following simple design heuristic:

1. Create a *nondimensional planform definition* based on the design brief – that is, on class/type of aircraft, range, Mach number – as well as on manufacturing/cost considerations.
2. Obtain initial guess at a target *wing area* S_{target} from constraint analysis based on the performance elements of the design brief; that is, take-off distances, rates of climb, and so on.
3. Create *geometry* by scaling the result of step 1 such that the area will equal the target value obtained from step 2 above.

This is a simple means of obtaining a geometry that satisfies the constraints, at least as far as the accuracy of the conceptual-level analysis can be trusted. The (parametric) geometry resulting from step 3 can then be placed at the centre of an iterative optimization process, which is guided by the same constraints as those we had used for the initial sizing, except that the analysis underpinning it can now benefit from the existence of a geometry (around which we can run, for instance, flow simulations).

As a simple example, let us consider the case when the planform sketch is that of a wing with a straight leading edge and a straight trailing edge. The *shape* of the planform (i.e. the nondimensional sketch) can be defined here unequivocally by these two geometrical statements plus aspect ratio AR, the taper ratio λ and a sweep angle – say, the leading edge sweep Λ_{LE}.

If we wanted, at this point, to visualize this shape, we could draw a unit root chord $C_{\text{root}} = 1$ and sketch the wing around it. This will almost certainly not have the S_{target} we obtain via step 2 above, but it will give us a starting point.

Now, with the taper ratio $\lambda = C_{\text{tip}}/C_{\text{root}}$ and the aspect ratio $\text{AR} = b^2/S$, the area of the (double) trapezoidal wing can be written as

$$S(C_{\text{root}}) = 2\frac{C_{\text{root}} + C_{\text{tip}}}{2}\frac{b}{2} = C_{\text{root}}^2 \frac{(1+\lambda)^2}{4}\text{AR}. \tag{8.16}$$

Step 3 of the heuristic outlined above now looks like an optimization problem: we need to find the value of C_{root} that gets us as close as possible to the desired target area S_{target}. More formally, this can be written as seeking to minimize the residual

$$R = \min_{C_{\text{root}}} |S(C_{\text{root}}, \lambda, \text{AR}, \Lambda_{\text{LE}}) - S_{\text{target}}|. \tag{8.17}$$

Of course, solving Equation 8.17 for a given λ and AR (and Λ_{LE}, though for this simple case the area is actually independent of it) is a trivial exercise here: the root chord that will yield the desired target area (with $R = 0$), from Equation 8.16, is

$$C_{\text{root}} = \frac{2}{(1 + \lambda)\sqrt{\text{AR}}} \sqrt{S_{\text{target}}}. \tag{8.18}$$

Nonetheless, while as far as optimization problems go this is not a very exciting one, it shows how the root chord can be used as a scaling control.

We hinted above that a nondimensional design could still be 'drawn' if we assumed some initial value for the root chord, and this is what we would do in a computational geometry context too: we would build a geometry around, say, $C_{\text{root}} = 1$ and then apply a scaling method to this 'unit' geometry. Here, by scaling we simply mean choosing a datum (say, the apex of the wing; that is, the intersection of the root chord and the leading edge) and *stretching* it along the three Cartesian axes such that all x, y, z dimensions are multiplied by the `ScaleFactor` (to introduce here the notation we will be using in the code snippets that follow). In this case the job is a simple one, as the number obtained from Equation 8.18 is also the stretching factor we would have to apply in all three dimensions.

8.5.3   More Complex Planforms and an Additional Scaling Factor

What if the area cannot be calculated as easily as with Equation 8.16 in the case of a trapezoidal wing? Some planforms can be broken down into multiple trapezoids, but analytical area calculations may become excessively tedious for some leading and trailing edge geometries (especially if the functional form of the definition of, say, the spanwise variation of the sweep angle changes in the course of the design process). Of course, an optimization problem similar to Equation 8.17 can still be written, but, in these cases, a numerical solution might be necessary.

Another common issue that may arise here is to do with the `ScaleFactor` introduced above. Perhaps the best way of viewing this number is to imagine moving the reference point into the Cartesian origin and then multiplying the coordinates of every point on the geometry by the `ScaleFactor` – this, of course, preserves the shape, and it is how all CAD engines define their basic 3D scaling method. However, what if we wanted to apply different magnitudes of stretching along the various axes?

What are the possibilities here? Would it be worth having, in fact, three separate scale factors (thinking ahead toward 3D wings, as we approach the end of this chapter on planforms)? In many areas of engineering the answer would be 'yes'. However, from a computational aerodynamic design perspective, would we really want to distort, for example, the shape of an aerofoil section by applying different scaling factors to the x and z coordinates? The answer is 'no'; the definition of the aerofoil section is best left to a separate method and, as once we have started assembling the 2D elements of the wing geometry, we should leave

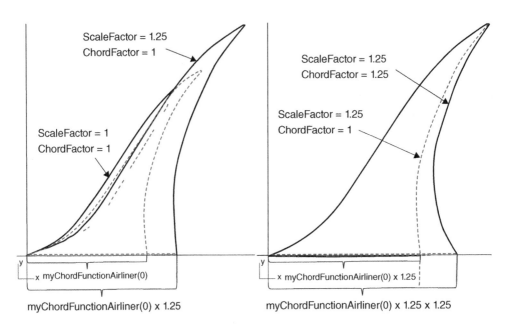

Figure 8.18 Illustration of the definitions of the two scaling factors we shall use in our objects instantiated from the LiftingSurface class. myChordFunctionAirliner is a user-defined function describing the spanwise variation of the chord length, as explained in Section 8.5.4 – it is supplied as one of the illustrative examples to allow the reader to test the capabilities of the LiftingSurface class (see Listing 8.1 for its definition).

these as they are. Should we really want to warp the aerofoil we could do this in its own 2D method, but playing with its shape there, as well as at the level of the complete wing, is an unnecessarily untidy and confusing manoeuvre (parameterizations based on freeform deformation, as discussed in Section 4.4, must be handled with particular care from this point of view – while there is a certain appeal to FFD's ability to change planform shape, aerofoil shape and scale all at once, the risks must be understood before embarking on aggressive geometry manipulations).

Here is an alternative the reader may wish to consider, and one which we have implemented in the LiftingSurface OpenNURBS/Rhino-Python class included with this book. We keep ScaleFactor as the standard, uniform 3D ratio of stretched and original coordinates, simply because of its intuitive nature and because it does exactly what a standard design process demands. For additional freedom, however, we also define a ChordFactor, which stretches the chord of each 2D aerofoil section we use to build the wing, while maintaining its proportions – in other words, it is basically a uniform x, z stretch applied to each section. See Figure 8.18 as an illustration of what this means in the context of the subject of this chapter; that is, planform design.

We can thus write 2D aerofoil definition methods that build unit chord aerofoils and pass them on to the wing definition method, which, in turn, will stretch them as dictated by the ChordFactor.

Listing 8.1   Chord length variation, as a function of ε, for the airliner wing example.

```
1  def myChordFunctionAirliner(Epsilon):
2
3      ChordLengths = [0.4724, 0.3792, 0.2867,
4      0.2272, 0.1763, 0.1393, 0.1155, 0.093,
5      0.0713, 0.0461, 0.007]
6
7      EpsArray = []
8
9      for i in range(0, 11):
10         list.append(EpsArray, float(i)/10)
11
12     f = linear_interpolation(EpsArray, ChordLengths)
13
14     return f(Epsilon)
```

8.5.4   Spanwise Chord Variation

Of course, with this logic we could only ever generate constant chord-length wings. There is, therefore, one more operation we have to apply to an aerofoil section before lifting it into the wing geometry: another 2D scaling step like that dictated by the ChordFactor, but this time with the possibility of varying the x, z scalings of the sections according to a spanwise law. Let us call this law ChordFunct and let us define it in terms of the spanwise parameter ε (we will define this properly in Chapter 9, but, for now, let us use the working definition that this is a dimension whose coordinate axis is the wing leading edge). For every spanwise station where we place an aerofoil section (more on how we do this in Chapter 9) we therefore scale that aerofoil with the product of ChordFactor and ChordFunct(Epsilon); that is, the local scaling factor at that particular ε station, as computed by the ChordFunct (Epsilon) method.

Listing 8.1 is the definition of the chord variation function we used to build the airliner wing geometry shown in Figure 8.18. While one of the beauties of separating these spanwise variations of various parameters into separate functions is that these functions can be richly parameterized and this does not increase the overall complexity of building the wing geometry (more on this later), we have based this particular wing definition on a fixed chord variation (to approximately match the planform of the Boeing 787-8 wing).

myChordFunctionAirliner simply returns a chord length (in the context of our wing construction process this is effectively an aerofoil section x, z scale factor) based on interpolating between 11 points. These chords will be attached to the leading edge and the trailing edge in a smooth interpolant of their end points – hence the elegant curved shape seen in Figure 8.18.

The spanwise variation of the 'local' sweep angle $\Lambda(\varepsilon)$ is defined in mySweepAngleFunctionAirliner(Epsilon) (included as one of the examples with the LiftingSurface class), which looks very similar to myChordFunctionAirliner and so is not reproduced here. Once again, for simplicity, this is a fixed function, though the scope for parameterization and flexibility enhancement is endless.

Listing 8.2   Python script for generating the airliner wing geometry using the LiftingSurface class.

```
 1  P = (0,0,0)
 2  LooseSurf = 1
 3  SegmentNo = 10
 4
 5  Wing = LiftingSurface(P, mySweepAngleFunctionAirliner,
 6   myDihedralFunctionAirliner, myTwistFunctionAirliner,
 7    myChordFunctionAirliner, myAerofoilFunctionAirliner,
 8     LooseSurf, SegmentNo)
 9
10  ChordFactor = 1.25
11  ScaleFactor = 1.25
12
13  Wing.GenerateLiftingSurface(ChordFactor, ScaleFactor)
```

In the latter half of this chapter we have started referring to our OpenNURBS/Rhino-Python class that exemplifies some of the wing geometry modelling concepts discussed in this book: LiftingSurface. While we have not discussed the 3D aspects of wing design and we have focused in this chapter on planform definition,[5] this is perhaps the right time for the reader to start experimenting with using LiftingSurface.

Listing 8.2 shows the typical wing construction process using LiftingSurface. The first line defines the location of the apex of the wing. This is followed by a definition of the type of surface fitting required (OpenNURBS/Rhino-Python offers the option of allowing a lofted surface, such as that of our wing, to regress – that is, deviate slightly from the supports in the interest of a smoother result) and the number of leading edge segments (and thus aerofoil sections).

Next, the object Wing is instantiated, with the two methods discussed above amongst the arguments (more on the other arguments in Chapter 9). Up to this point we have a mathematical definition of a dimensionless entity, but no actual model yet.

For that, we next define the scale factors discussed earlier, corresponding to the largest of the planforms shown in Figure 8.18. With these we can now call the actual geometry building method GenerateLiftingSurface, applied to the newly created object Wing. Note that we could build other Wings with different scale factors – the object is simply the definition of the *shape*.

Of course, the reader may wish to implement a similar class in a different environment, and there is no impediment to this – none of the software engineering philosophy implemented here is specific to *Python, OpenNURBS* or *Rhino*.

An aircraft wing is a 3D object and, as manufacturing constraints are gradually lifted (for example, by technologies like additive manufacturing), this 'three-dimensionality' (to use an ungainly and fuzzy, but intuitive term) is increasingly exploited – we have progressed from the early 'flying barn doors' to wings with complicated 3D surfaces. The aerodynamics of

---

[5] This is a slight abuse of the term 'planform', which is actually a projection of a wing on a horizontal plane and, therefore, is influenced by 3D aspects, such as twist and dihedral too; but the latter, in most cases, only have second-order effects on planform shape.

many wings exhibits strong 3D effects too. However, these can be tricky to grasp directly, and aerodynamic design engineers like to first break down a simplified version of the problem to its 2D component parts.

This thinking has been traditionally followed in geometry definition too, partly for the reasons mentioned in the introduction to this chapter. In Chapters 5, 6 and 7 we looked at 2D aerofoil sections, and in this chapter we tackled the description of 2D wing planforms. It is now time to put these into their 3D places through a process we shall term the *synthesis* of the geometry of a wing.

9

Three-Dimensional Wing Synthesis

The airflow patterns around most wings of any practical significance are clearly 3D. Yet, the advent of sophisticated 3D modelling and visualization tools has not changed the fundamental fact that the human mind instinctively attempts to divide and conquer 3D physics and geometry by attempting to understand its 2D projections first. This is, of course, also linked to the physical expediency of picking up pencil and paper and sketching – in spite of some radical hardware innovations in the pipeline (3D pens), this is set to remain a largely 2D activity for some time yet.

The divide and conquer principle does help immensely when constructing parametric geometry models too, which is why we discussed the two key 2D geometries – aerofoil section and planform – in some detail in the previous chapters. The time has come now, however, to build a 3D geometry upon these foundations. As a first step, we discuss two 'classic' design parameters that offer a convenient basis for the construction of a 3D wing.

9.1 Fundamental Variables

9.1.1 Twist

We have already encountered a number of seemingly innocuous geometrical terms loaded with subtle complexities, and wing twist is yet another. The term refers to the spanwise variation in the 'pitch' orientation of wing sections – that is, the rotation of the aerofoil sections in their plane of definition, but there are two possible interpretations here.

There is a *geometrical twist*, which describes the variation in the setting angles of the wing section chords. This is the most commonly used sense of the term. There is a related concept, *aerodynamic twist*, which describes the variation in the setting angles of the sections relative to their respective zero-lift lines. The difference between the two may be worth considering when the shape of the aerofoil changes significantly in the spanwise direction.

A wing featuring sections in a more 'nose down' position at the tip than at the root is said to have *washout*, with the term sometimes also used to denote the related angle (the difference

Aircraft Aerodynamic Design: Geometry and Optimization, First Edition. András Sóbester and Alexander I J Forrester.
© 2015 John Wiley & Sons, Ltd. Published 2015 by John Wiley & Sons, Ltd.

between the root and the tip twist). The opposite is *wash-in*, though this is a relative rarity (we mentioned an example in Section 8.3.4).

There are two main reasons why wings may need to be twisted:

- *Stall behaviour.* The aerofoil sections of an untwisted wing will reach their stall angles of attack at the same time, thus causing a complete and abrupt stall of the entire wing. This may not be a problem (e.g. in the case of aircraft equipped with an envelope protection system that prevents them from getting to that angle in the first place), but it is generally considered good practice to engineer a stall that spreads progressively along the wings, root first. This gives the pilot time to react after the early indications of the onset of root stall (e.g. buffeting), before lift is lost completely. With roll control surfaces usually near the wingtips (such positioning maximizes their authority through their longer moment arm), maintaining attached flow over the outboard sections of the wing for as long as possible is generally considered a benign handling trait.
- *Load tailoring.* Engineering a particular lift distribution profile along the wing is of interest both from aerodynamic and structural engineering points of view. By varying the twist along the wing, different lift distributions can be obtained, giving varying levels of induced drag and/or structural efficiency.

Some of the effects of twist can, in general, be achieved in other ways; for example, by using aerofoils with lift coefficients that vary with the span or by contouring the planform. Both of these measures, however, raise manufacturing issues that are more expensive to solve than twist. At least, they are more expensive than *linear* twist – a nonlinear twist variation is generally difficult to achieve in practice. Nevertheless, modern manufacturing techniques – in particular, various additive manufacturing methods – are likely to eliminate this drawback and may permit an optimization algorithm to search a broad design space space of twist distributions. In this spirit, when we refer to 'twist' in what follows, we mean a *local twist angle*, as a function of some spanwise coordinate, just as we talked of a local leading edge sweep angle in Section 8.3.1 and as we will talk of a local dihedral angle next.

9.1.2 Dihedral

Structural and manufacturing reasons generally favour a planar wing. Most importantly, this allows an efficient support of the air loads through a straight main spar running from wingtip to wingtip.

Twist, as we saw in Section 9.1.1, is a deviation from this ideal in the 'pitch' direction; that is, a rotation of the wing aerofoil sections around the y-axis. If aerofoil sections perpendicular to the surface of the wing are rotated around their *chord* (or the x-axis) and the surface also rotates such that it remains perpendicular to the section, the wing is said to feature *dihedral* if the rotation is such that a point immediately outboard from the reference point is higher above the ground than the reference point. If a point immediately outboard of the reference section is below it, the wing is said to have *anhedral*.

The structural sacrifice necessitated by the introduction of dihedral or anhedral is usually warranted by considerations of lateral stability – more specifically, to increase the so-called *dihedral effect* linked to the rolling moment due to sideslip. Dihedral increases this stability derivative, as does sweep, as does having a high wing, far above the centre of gravity of the aircraft. In fact, if the latter two are present, dihedral is not only unnecessary, but often a pronounced anhedral is needed to *weaken* the combined dihedral effect of these factors – see Figure 9.1 for examples of aircraft with high, swept wings with anhedral.

In the vast majority of cases the anhedral angle is constant along the span, with a few exceptions where partway along the span a change from planar to dihedral or anhedral to dihedral ('gull wing') occurs – see Figure 9.2 for examples of such aircraft. These (usually low wing) aircraft require a more pronounced dihedral effect than a planar wing would give them – hence the dihedral on the outboard section – but there are unrelated reasons for needing an anhedral or a planar section near the fuselage; for example, to reduce the wing/fuselage junction interference drag by ensuring that the plane of the wing is normal to the fuselage surface.

The comparative rarity of aircraft featuring polyhedral geometry wings is due largely to the substantial weight, cost and complexity penalties often associated with breaking up a wing in this way – consider, for instance, the complicated multisegment flap system imposed upon the designers of the Vought Corsair (see Figure 9.3) by the change from anhedral to dihedral partway along the span. Such elaborate systems are generally the tell-tale sign of a complex trade-off analysis process, often involving multiple disciplines and high-level mission analysis (at the most basic level, does the added weight make up for the aerodynamic and/or handling benefits of the gull wing design?).

Multiple changes of the dihedral/anhedral angle along the span are rare, and nonlinear changes especially so. However, from a geometry modelling standpoint, following the pattern of previously discussed wing parameters, it is worth introducing the concept of a *local dihedral angle* $\Gamma(\varepsilon)$ (where ε is a spanwise coordinate axis). There are a number of reasons for this:

- With developments in manufacturing technologies it will eventually become possible to prescribe any (continuous) spanwise variation of Γ. In fact, this is already possible in the case of small, unmanned aircraft, in particular when using additive manufacturing technologies – see Figure 9.4 for an example, where Γ varies from 0 to 180°. We shall look at the construction of this geometry in detail in Section 9.3.1.
- The geometrical description of blended winglets is greatly facilitated by viewing them as a wing with variable Γ – we look into this in detail in Section 9.3.2.
- The software integration of fluid–structure interactions studies can be streamlined by being able to use the same parametric geometry formulation that was used to engineer the unloaded geometry of a wing – see Figure 9.5 for an example.

With the two 2D facets (aerofoil sections and the planform) and the two key '3D' wing design variables (twist and dihedral) ready we are now almost prepared to assemble all the pieces to obtain the final, 3D parametric wing geometry. All we need now is a 'scaffold' to build it upon. This is what we address next.

9.2 Coordinate Systems

9.2.1 Cartesian Systems

The overwhelming majority of aircraft geometry description formulations are based on Cartesian coordinate systems. A typical set-up comprises a set of slave systems defining components

Figure 9.1 Aircraft with high, swept wings, featuring pronounced anhedral. Clockwise from top left: Harrier (AV8-B), Airbus A400M, British Aerospace BAe-146, Ilyushin Il-76 (photographs by A. Sóbester).

Figure 9.2   Aircraft with polyhedral wings. Anticlockwise from top: Jodel D18, Vought F4U-1D Corsair and the Mahoney Sorceress, a staggered biplane designed for the Reno Air Races (photographs by A. Sóbester).

Figure 9.3   Multisegment flaps on the polyhedral wing of a Vought Corsair (photograph by A. Sóbester).

Figure 9.4 CAD rendering of a box wing unmanned aircraft – an example of $\Gamma(\varepsilon)$ varying linearly between the end of the zero dihedral lower half of the wing and an upper half that can be viewed, from a purely geometrical standpoint, as having 180° dihedral.

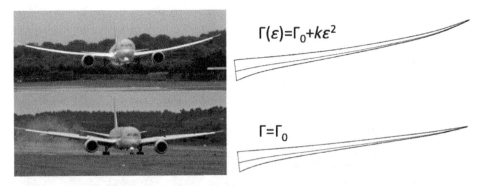

Figure 9.5 Local dihedral angle variation defined along a wing-attached, spanwise coordinate axis ε, to model deformed shapes for fluid–structures interactions studies (Boeing 787-8; photographs by A. Sóbester).

(wings, fuselage, engines, landing gear, etc.) and subcomponents (e.g. control surfaces) and a master system used to define the relative positions of the components.

Most aircraft have a central, vertical symmetry plane,[1] and this serves as a convenient datum to which to link the master coordinate system – specifically the xOz plane of the master coordinate system, with the x axis pointing downstream and the z axis pointing upwards. The standard right-handed system is completed with a y-axis, whose positive direction is outboard.

---

[1] Of course, as the design process progresses and more detail is added the strict symmetry is lost, but the original 'conceptual' symmetry plane still remains a natural reference.

The larger, symmetrical subcomponents generally inherit the orientation of the master system, and this is the case with the wings too, though it is common practice to translate the origin to the intersection of the leading edge and the central symmetry plane. This coordinate system then serves as a master series of slave systems, which define the aerofoil sections of the wing.

This is conceptually rather straightforward, but it suffers from one major drawback: the mathematical description of the wing becomes increasingly cumbersome as more 'exotic' wing shapes are desired. For instance, consider describing in terms of x, y and z the spatial orientation of the slave Cartesian system attached to the tip aerofoil of a wing featuring a non-linear twist distribution, a curved leading edge and a blended winglet or perhaps a wing like that seen in Figure 9.4. Ultimately, a series of multiplications by transformation matrices make this possible, but the resulting algebraic form will make little intuitive sense and, most importantly, will make sensible parameterization all but impossible (the impact of changing any parameters you may choose will be lost in a fog of multiple coordinate system rotations and translations). Let us consider a more natural alternative.

9.2.2   A Wing-Bound, Curvilinear Dimension

Consider a parametric wing, the aerofoil section of which varies in a spanwise direction. Let us imagine that not only does its scale change (as in the rather common case of a tapered wing), but, say, its camber too. There could be a number of reasons for wishing to do this: for instance, to achieve better control of the spanwise lift distribution. In the case of simple wings it might be convenient to define this variation in terms of y; but what if the wing is swept, with a curved leading edge and with nonlinear dihedral, perhaps ending in a vertical segment (such as the top part of a blended winglet)? Mathematical complexities may result from the former features, but the latter would make y entirely unsuitable – it will be constant for the vertical component. What if the wing is to 'fold back onto itself', as in the case of a box wing?

Messy workarounds (such as breaking the wing down into segments) could be used, but a cleaner solution is that illustrated in Figure 9.6. Consider an extra dimension – let us call it ε – whose curvilinear axis is 'glued' to the leading edge and whose scale is normalized, such that at the wing root $\varepsilon = 0$ and at the tip (or the 'other root', as in the case of the box wing shown on the right of Figure 9.6) $\varepsilon = 1$.

Defining the spanwise variation of camber – or, indeed, the spanwise variation of anything – in terms of ε is now trivial, whether we are dealing with a plane rectangular wing, a transonic airliner wing (Figure 9.6, left), a box wing, a gull wing (see lower half of Figure 9.2 for examples) or any other less common geometry. In fact, even complete loops, as on a spiroid wingtip or the cowling of a turbofan engine can be defined in terms of the curvilinear, leading edge bound coordinate without any further mathematical artifices.

9.3   The Synthesis of a Nondimensional Wing

There are a number of ways in which a wing geometry can be generated. The choice depends to some extent on the type of software environment in which the geometry will be implemented, as well as on the envisaged flexibility. Here, we aim for a method that is relatively simple

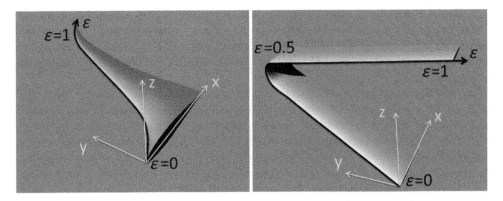

Figure 9.6 Two wings (transonic transport on the left, box wing on the right) illustrating a conventionally orientated Cartesian system with its origin at the root of the leading edge and an additional curvilinear dimension ε attached to the leading edge.

to implement in a CAD-type environment, as well as being in keeping with our overarching desire for parameterization flexibility.

In the spirit of the 'divide and conquer' principle we have already invoked a number of times, we first define the key geometrical features of the wing in isolation, which can then be readily synthesized into a complete, unequivocally defined wing model. These features are:

- *Aerofoil cross-section*, as a function of ε, built on a unit chord.
- *Twist* variation, as a function of ε. Measured at the leading edge.
- *Chord* variation, as a function of ε.
- Local *dihedral angle* variation, as a function of ε.
- Local *sweep angle* variation, as a function of ε.

Synthesizing the wing then amounts to the following steps:

1. Define the number of spanwise stations where aerofoil sections will be constructed. This will yield an inter-aerofoil step length.
2. Starting from the root, position the leading edge points of each aerofoil in space by taking discrete steps of the length established in step 1 in a direction determined in the horizontal plane by the local *sweep angle* and in the vertical plane by the local *dihedral angle* (note that the ε-wise variation of both can be highly nonlinear, e.g. on a blended winglet).
3. At each of these points construct an aerofoil scaled to the local *chord length*.
4. Rotate each aerofoil around its leading edge (in a plane parallel with xOz) by the local *twist angle*.
5. Rotate each aerofoil around its chord by the local *dihedral angle*, such that they will be orthogonal to the leading edge.
6. Loft a surface over the aerofoils thus generated. Depending on whether exact matching of the aerofoils or surface smoothness is more important, this can be an interpolating surface or a regression surface respectively.

Listing 9.1   Aerofoil definition in OpenNURBS/Rhino-Python as a function of ε – constant NACA 5310 throughout the span. The aerofoil is constructed here as an instance of the `Aerofoil` class introduced in Chapter 5.

```
 1  def myAerofoilFunctionBoxWing(Epsilon, LEPoint, ChordFunct,
 2                           ChordOffset, DihedralFunct, TwistFunct):
 3
 4      AerofoilChordLength = (ChordOffset+ChordFunct(Epsilon))/
 5                      math.cos(math.radians(TwistFunct(Epsilon)))
 6
 7      Af = primitives.Aerofoil(LEPoint,
 8                          AerofoilChordLength,
 9                          DihedralFunct(Epsilon),
10                          TwistFunct(Epsilon))
11
12      SmoothingPasses = 1
13
14      Airf, Chrd =
15      primitives.Aerofoil.AddNACA4(Af, 5, 3, 10, SmoothingPasses)
16
17      return Airf, Chrd
```

9.3.1   *Example: A Blended Box Wing*

Let us consider an example of wing synthesis: the geometry of a relatively simple box wing. To begin with, let us define a simple, baseline wing, with zero sweep and dihedral, constant chord and a NACA 5310 aerofoil section.

The 'aerofoil as a function of ε' definition is clearly trivial in this case, though Listing 9.1 of our OpenNURBS/Rhino-Python implementation does include some data handling steps, which we have included here for completeness. Essentially, on line 7 we define `Af` as an object of class `primitives.Aerofoil` and on line 14 we call the method `AddNACA4` now associated with the object to actually generate a NACA 5310. The object needs to be equipped with some other attributes needed to specify the location and orientation of the aerofoil, such as the leading edge point and the handles of the other functions of ε that will define the rest of the geometry, but all that can be simply regarded as part of a template – the user should only really change the section between lines 12 and 15. Figure 9.7 is a sketch of the resulting wing (assuming that the sweep and dihedral functions return zero for any ε).

Let us now fold this wing back onto itself. This can be easily achieved by starting off with a local dihedral angle Γ of $0°$ and transitioning to a dihedral of $180°$ somewhere along the way (we are stretching the concept of 'dihedral' here somewhat, but the meaning should be relatively intuitive). A simple and smooth way of achieving the transition is a linear one. Starting at, say, 45% of the leading edge length and using, say, 10% of the leading edge length to complete it, we can define the local dihedral angle as

$$\Gamma(\varepsilon) = \begin{cases} 0 & \text{if } \varepsilon < 0.45 \\ 180\frac{\varepsilon-0.45}{0.1} & \text{if } \varepsilon \in [0.45, 0.55] \\ 180 & \text{if } \varepsilon > 0.55 \end{cases} \tag{9.1}$$

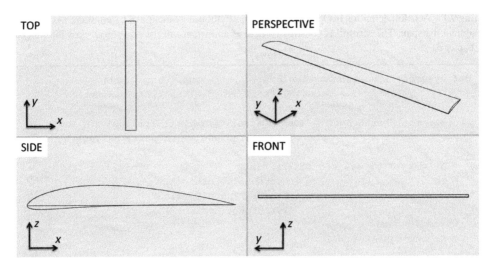

Figure 9.7 The first step of building the box wing: a simple straight, plane wing with a NACA 5310 section.

Listing 9.2 shows a slightly more generic version of this as an input to our OpenNURBS/Rhino-Python parametric wing geometry.

Figure 9.8 is a sketch of the resulting wing and a hint of the next step of the process. The folding process has obviously yielded a wing that will not generate lift (or, at least, not in an efficient manner), as the direction of the camber changes with Γ. A simple fix is to impose a gradual 'flip' on the camber in a manner similar to the local dihedral angle variation of Equation 9.1.

Listing 9.3 shows how this can be accomplished in the OpenNURBS/Rhino-Python implementation of the parametric wing geometry described here and Figure 9.9 is a sketch of the

Listing 9.2 Box wing dihedral definition as a function of ε.

```
 1  def myDihedralFunctionBoxWing(Epsilon):
 2      D1 = 0
 3      D2 = 180
 4      Transition1 = 0.45
 5      Transition2 = 0.55
 6
 7      if Epsilon < Transition1:
 8          return D1
 9      elif Epsilon > Transition2:
10          return D2
11      else:
12          return D1 + ((Epsilon - Transition1)/
13                       (Transition2 - Transition1))*(D2-D1)
```

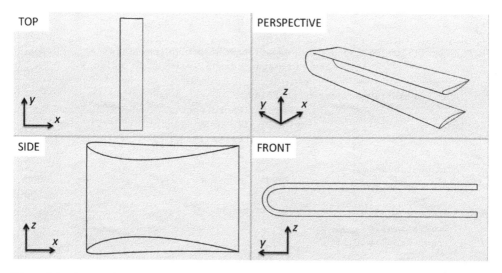

Figure 9.8 Step 2 of generating a box wing: the basic wing is folded back onto itself through a linear transition.

new geometry. Incidentally, all these tweaks also suggest that the divide and conquer philosophy we followed here, as well as the use of functionals as inputs to the wing geometry, make increasing the flexibility of the geometry very simple. For instance, other camber variation schemes – perhaps designed to tailor the lift distribution of the wing – could be integrated quite readily. The aerofoil function code of Listing 9.3 allows 'out of the box' variation of the camber, the streamwise location of the maximum camber and the maximum-thickness-to-chord ratio (the input parameters of the NACA four-digit foil implemented there), but the `primitives.Aerofoil.AddNACA4` method therein could be replaced quite readily with a different, perhaps more (or less) flexibly parameterized aerofoil, the parameters of which could vary with `Epsilon` in the same way as `Camber` does in this example.

Finally, let us consider adding a constant sweep to both elements of this box wing, such that the planform opens up into an 'A' shape (see Figure 9.10). Once again, courtesy of the ε curvilinear coordinate axis, this is a very straightforward operation, as seen in Listing 9.4.

This is, of course, as defined here, a hard-coded, nonparameterized geometry. However, there are abundant opportunities to add flexibility here. Take, for example, the sweep component of the geometry definition: `mySweepAngleFunctionBoxWing` (Listing 9.4) has some obvious candidates for parameterization – `S1`, `S2`, `Boundary1` and `Boundary2` could be used as design variables. There is, however, so much more we could do in this object-oriented framework: having these simple user-defined functions as arguments to the wing definition means that their functional form could be varied too. This means that the wing geometry can be reparameterized one aspect (such as sweep variation) at a time, but also that we could even build an optimization process capable of sweeping not only a range of parameter values, but also *a range of parameterizations*.

Listing 9.3 Aerofoil definition as a function of ε: camber smoothly 'flipped' at the folding point.

```
1  def myAerofoilFunctionBoxWing(Epsilon, LEPoint, ChordFunct,
2                              ChordOffset, DihedralFunct, TwistFunct):
3
4      AerofoilChordLength = (ChordOffset+ChordFunct(Epsilon))/
5                          math.cos(math.radians(TwistFunct(Epsilon)))
6
7      Af = primitives.Aerofoil(LEPoint,
8                          AerofoilChordLength,
9                          DihedralFunct(Epsilon),
10                         TwistFunct(Epsilon))
11
12     SmoothingPasses = 1
13
14     Camber1 = 5.0
15     Camber2 = -5.0
16     Transition1 = 0.45
17     Transition2 = 0.55
18
19     if Epsilon < Transition1:
20         Camber = Camber1
21     elif Epsilon > Transition2:
22         Camber = Camber2
23     else:
24         Camber =  Camber1 + ((Epsilon - Transition1)/
25                 (Transition2 - Transition1))*(Camber2-Camber1)
26
27     Airf, Chrd =
28     primitives.Aerofoil.AddNACA4(Af, Camber, 3, 10, SmoothingPasses)
29
30     return Airf, Chrd
```

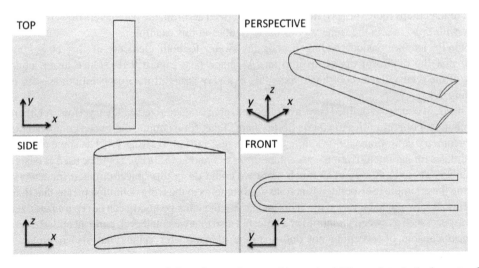

Figure 9.9 Folded wing with aerofoil section camber transition at the folding point – both elements of the wing have positive camber.

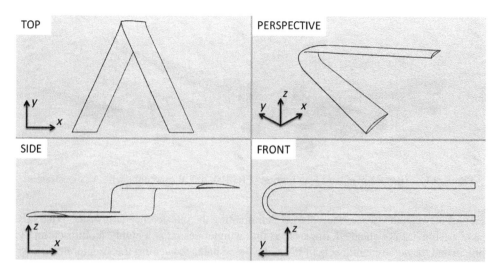

Figure 9.10   Completed box wing.

9.3.2   Example: Parameterization of a Blended Winglet

The aerodynamic principle behind most wingtip devices is the reduction of induced drag. These have tended to be additional lifting surfaces (or flat plates) attached to the main wing at some cant angle, with a relatively sharp transition between the two. This has the disadvantage of being a source of interference drag, which may cancel out some of the induced drag gains.

The latest generation of winglets, designed mostly as retro-fit bolt-ons to existing wings, but increasingly as part of new designs too, feature a gradual transition (see Figure 9.11 for an

Listing 9.4   The final step in building the box wing: defining the sweep variation.

```
 1  def mySweepAngleFunctionBoxWing(Epsilon):
 2      # User-defined function describing the variation of sweep angle
 3      # as a function of the leading edge coordinate
 4
 5      S1 = 25
 6      S2 = -25
 7      Boundary1 = 0.45
 8      Boundary2 = 0.55
 9
10      if Epsilon < Boundary1:
11          return S1
12      elif Epsilon > Boundary2:
13          return S2
14      else:
15          return S1 +
16              ((Epsilon - Boundary1)/(Boundary2 - Boundary1))*(S2-S1)
```

Figure 9.11   Blended winglets on an Embraer ERJ 190-200LR (photograph by A. Sóbester).

example) between the two surfaces, largely eliminating the interference drag issue. What this type of design has not changed, however, is that winglet design is a highly multidisciplinary problem and the aerodynamic benefit must always be balanced against the structural and cost implications (Ning and Kroo, 2010).

The prerequisite of any such study is, of course, a parametric geometry. Let us consider here a very simple model that provides us just that. Consider Figure 9.12. It depicts two one-dimensional parameter sweeps along the variables of a parametric blended wing geometry. One of the variables (six instances with different values of this variable are shown on the left) controls the slope of the tangent at the tip; that is, the dihedral angle at the wingtip (one might also term this the *cant angle* of the winglet).

The same geometry, with the tip tangent slope set at the highest of the values instantiated on the left, is shown on the right of the figure, this time in six instances, each with a different value of the other design variable: the spanwise station where the transition from the baseline dihedral to the specified wingtip tangent (the blend) begins.

In the wing geometry parameterization framework described here, the description of the parametric winglet is thus simply a particular variation of the local dihedral angle near the tip, and thus it can be integrated into the dihedral function – see Listing 9.5 (showing one of these instances – of course, for the purposes of an optimization study, `WingletStart` and `WingtipTangent` would be variables passed to the function by an optimizer).

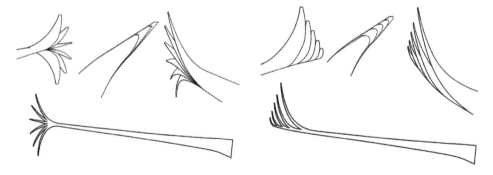

Figure 9.12   A simple, two-variable parameterization of a blended winglet geometry. Sweeps of the variable controlling the tip tangent (left) and the variable controlling the transition point (right).

Listing 9.5 A two-variable parametric blended winglet model.

```
1  def myDihedralFunctionAirliner(Epsilon):
2      BaseDihedral = 6
3      # Design variable one
4      WingletStart = 0.75;
5      # Design variable two
6      WingtipTangent = 75;
7
8      if Epsilon < WingletStart:
9          return BaseDihedral
10     else:
11         return BaseDihedral +
12         WingtipTangent *
13         (Epsilon - WingletStart)/(1 - WingletStart)
```

This simple parametric winglet has one more feature: multiple instances of it can be combined (through a standard union operation), allowing the construction of more complex wingtip devices, such as the *scimitar winglet* shown in Figure 9.13.

9.4 Wing Geometry Scaling. A Case Study: Design of a Commuter Airliner Wing

In the shape definition framework discussed here, following the principle of 'divide and conquer' we broke down the shape definition into easily tractable components, which we then synthesized into a nondimensional wing geometry – or, to be specific, a wing geometry of unit leading edge length (ε has a range of [0, 1]) and unit root chord.

This recipe is by no means the only computational process whereby a 3D wing shape can be built up. However, regardless of construction methodology, the next step is to scale the resulting shape, typically to a target projected area.

Let us now consider this process, illustrated through an example: the design of the wing of a commuter-class twin turboprop. Our starting point is a relatively simple geometry: a linearly

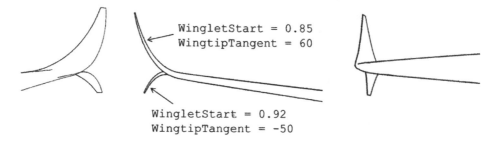

Figure 9.13 Scimitar winglet generated by combining two instances of the parametric blended winglet. The two component winglets differ on the starting point of their transition, the wingtip tangent, as well as on the overall scaling factor.

Table 9.1   Commuter turboprop wing – nondimensional parameters.

Variable	Value
Twist	−2° (washout)
Root aerofoil	NACA 63A418
Tip aerofoil	NACA 63A412
Aspect ratio	10.0
Taper ratio tip/centreline	0.333
Sweep on 30% chordline	0
Leading edge sweep	3°
Dihedral	7° (constant)
Take-off, first stage of flaps C_L^{max}	1.45
Final stage of flaps, up to warning C_L^{max}	1.98
Clean, up to warning C_L^{max}	1.63

twisted, 1/3-tapered wing with straight edges and an aspect ratio of 10. Table 9.1 contains a listing of the parameters of this nondimensional wing sketch, as well as some of the key scale-independent aerodynamic parameters.

Note the difference here between the various maximum lift coefficients – in particular, the maximum lift coefficient on take-off, as estimated for this geometry, being relatively low (taking the other two values as a reference). Relatively conservative C_L estimates are usually used for take-off to leave two key margins. The first one is based on a purely geometrical reasoning, which can be revisited in the latter stages of the design process: the maximum achievable angle of attack may be limited by a tail-strike constraint. The second is to do with the stall warning margin – the C_L achievable on take-off without stall warning and stick shaker activation may be two- or three-tenths lower than the ultimate value.

Such coefficients can usually be obtained at this stage of the design process from empirical models based on the performance of existing, similar designs or numerical simulations.[2]

If we wanted to generate a visual representation of this design – for instance, via our OpenNURBS/Rhino-Python code – we could simply build a wing with an aspect ratio of 10 and an arbitrary target area or an arbitrary root chord, remembering that, at this stage, we are looking at a scale-free, back-of-a-napkin sketch. Figure 9.14 shows this sketch.

To build an actual scaled version of a wing of this shape, we need to consider the relevant design constraints. These are largely derived from the mission definition of the aircraft (some requiring estimates of speeds and weights) and the relevant certification constraints. Let us assume that our turboprop is to satisfy the following set of constraints.

- *Initial climb:* 2.4% minimum climb gradient with gear down at $V_2 = 120$ KTAS after losing one engine (propeller considered in the position it rapidly and automatically assumes after

---

[2] Of course, aerodynamic derivatives, such as those listed there, are technically not scale independent: they may vary with Reynolds number and Mach number. Such variations, however, are insignificant within the range of sizes, speeds and altitudes typically associated with this class of aircraft; so, for practical purposes, they can be considered scale independent. CFD simulations can also be conducted at an approximately appropriate Reynolds–Mach number combination and the resulting nondimensional coefficients used on the assumption that these scale factors will not be too different from those corresponding to the final design.

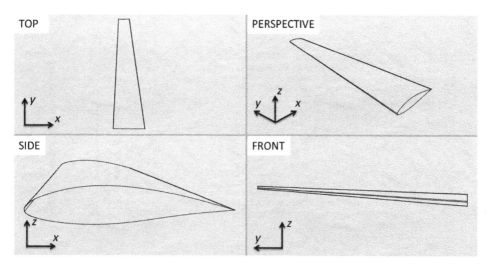

Figure 9.14 Commuter-class turboprop wing sketch.

loss of power, rudder deflected to counteract yawing moment) following take-off. Denver ISA normal conditions apply; that is, an airport elevation of 5500 feet (1676 m) and a temperature of 4°C at runway level.

- *Service ceiling:* achieve a climb rate of 500 ft/min at a density altitude of at least 7600 m. The en-route climb speed should be between 170 and 210 KTAS.
- *Take-off:* ground roll not exceeding 1900 m at ISA normal conditions on a runway; 50 ft obstacle to be cleared in 2000 m (both at Denver ISA normal conditions).
- *Target cruise speed:* 230 KTAS at the service ceiling.

Figure 9.15 is a graphical representation of these constraints in thrust-to-weight/wing loading space, computed using simple models along the lines of the discussion in Section 8.4.1.

In addition to the two rate of climb constraints (initial and service ceiling), the take-off distance constraints and the stall speed constraints, the figure also features a series of vertical lines showing the preferred wing loadings corresponding to various flight conditions and requirements, namely the wing loading values that maximize range and endurance for propeller-driven and jet aircraft respectively. The continuous lines correspond to the target cruise speed of 230 KTAS, while the thick dashed lines either side of these reflect the impact of altering the airspeed slightly (the maximum L/D corresponding to the increased cruise speed is not shown as it is very deep into the infeasible domain, out of the bounds of the graph).

Note the significant sensitivity of the locations of these lines to changes in airspeed – alterations of 10% either way in the true airspeed move the optimum wing loading by over 40 kg/m^2. In order to reduce clutter we did not include further speeds/altitudes and other perturbations, but the type of constraint analysis described here should typically be conducted with a close eye on the cruise/loiter efficiency implications of wing loading choices. If there is some flexibility in terms of the airspeeds to be achieved (this is usually the case with the best endurance speed in particular) the diagram can be read 'backwards'; that is, the highest

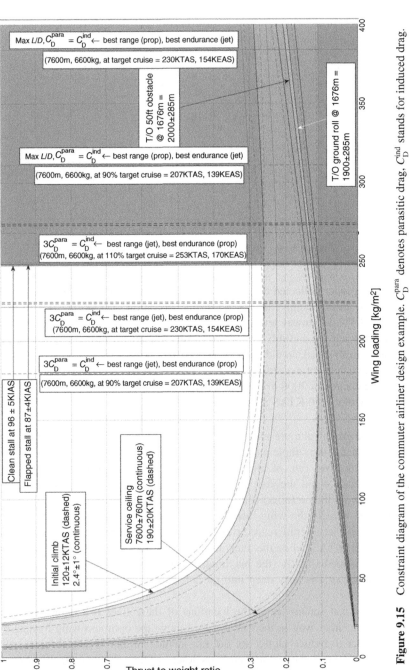

Figure 9.15 Constraint diagram of the commuter airliner design example. C_D^{para} denotes parasitic drag. C_D^{ind} stands for induced drag.

feasible wing loading can be found (as defined, say, by the stall constraint) and the speed can then be derived that optimizes cruise/loiter for that wing loading.

Figure 9.15 shows similar sensitivity analyses around the constraint boundaries too. Consider, for example, the initial climb constraint. The dashed lines either side of the boundary correspond to increasing (thick line) or reducing (thinner line) the initial climb speed V_2 by 10%. The graph also tells us that if we could make a 1° concession in the initial climb angle (continuous lines either side of the constraint boundary), we could get away with a lower thrust-to-weight ratio: by about 0.03 for a design with a wing loading of around 150–200 kg/m^2.

Stall speed requirements often define the right-hand edge of the feasible region, and this commuter turboprop design is no exception. The flapped and clean stall speed constraints almost completely overlap, forming a common boundary at around the 250 kg/m^2 mark. The sensitivity of this figure to the stall speed is remarkable and is exemplified by the 'ghost' boundaries (thin dashed lines) representing 5 knot and 4 knot deviations from the target clean and flapped stall speeds respectively. The magnitudes of these offsets are an indication of what a strong driver of wing area the stall speed constraint is.

The take-off ground roll constraint and the take-off and 50 ft obstacle clearance constraint near the bottom edge of the diagram are inactive in this particular case, as the initial climb constraint is more restrictive than either of them.

The Constraint Diagram as a Sizing Tool

The goal of a constraint analysis exercise such as the one shown above is to select two key preliminary design variables: the required engine size and the wing area.

In theory, once we have a constraint diagram, the algorithm is quite straightforward: select a design point in the feasible region and, based on an initial weight estimate, derive the required thrust and wing area from the thrust-to-weight ratio and the wing loading respectively.

The practical question, however, is where exactly in the feasible region do we place the design point? A good initial guess might be near the bottom right-hand corner of the feasible region; this will ensure that we have the smallest engine and the smallest wing we can get away with, both being generally considered as desirable from a weight and cost point of view.

However, this point may not be optimal from a cruise range point of view and/or from a loiter endurance point of view. It will almost certainly not be optimal on both scores, which also raises the interesting question of how much cruise efficiency are we willing to sacrifice in the interest of improved endurance (i.e. improved holding pattern performance).

All of this, of course, highlights the need for an iterative, multistage design process, where we constantly redraw the constraint diagram and redesign the putative aeroplane, perhaps using different points from within the feasible region.

A further complication is introduced by the fact that it is generally not possible to represent a design by a single point on the constraint diagram, as different constraints might be activated at different design conditions; that is, at different weights and different parts of the engine performance curves.

To illustrate this point in the context of the commuter turboprop example, let us consider a 'real' aircraft of this category: the BAe Jetstream 31 (we actually loosely based the geometry and the design brief of the example on this aeroplane). Table 9.2 summarizes some of the key features of this aircraft (an image of which is shown in Figure 9.16).

Table 9.2 Jetstream 31 – basic performance data.

Variable	Value
Propulsion	2× Garrett TPE 331 turboprops
Power	2×940 SHP
Maximum take-off weight (kg)	7059
Zero-lift drag C_{D_0}	0.038
Maximum lift-to-drag ratio L/D	10.8 (at $C_L = 0.8$)
Cruise speed (knots)	230
Maximum speed (knots)	260
Balanced field length – ISA 15°C, sea level (m)	1200
Balanced field length – ISA 20°C, 5000 ft/1524 m (m)	2440
Rate of climb – ISA 15°C, sea level (m/s)	8.6
Service ceiling (m)	7600
Landing field length – ISA 15°C, sea level (m)	760
Stall speed (clean) (knots)	96 (indicated)
Stall speed (35° flap) (knots)	87 (indicated)
$C_{L_{max}}$ (clean)	1.633
$C_{L_{max}}$ (35° flap)	1.980
Min. speed (knots)/max. C_L with no stall warning (clean)	109/1.320
Min. speed (knots)/max. C_L with no stall warning (35° flap)	98/1.633

Figure 9.17 shows the set of constraints of the design example of the previous section as a backdrop to the design points (red diamonds with red labels) representing the Jetstream 31 on a thrust-to-weight versus wing loading chart. We considered three distinct points here: sea-level take-off at a weight of 6600 kg, take-off at the same weight from Denver airport and at the service ceiling of 7600 m after having burnt off 600 kg of fuel.

As expected, these design points are largely clustered near the bottom right-hand corner of the feasible region, with the service ceiling design point sitting right on the service ceiling constraint boundary and the take-off design points within the 5% margin of the stall speed

Figure 9.16 BAe Jetstream 31, seen here in the livery of the UK's National Flying Laboratory (courtesy of G-NFLA).

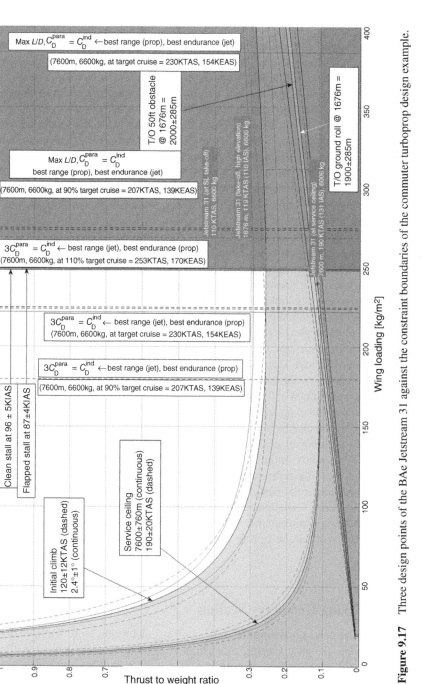

Figure 9.17 Three design points of the BAe Jetstream 31 against the constraint boundaries of the commuter turboprop design example.

boundaries. All three wing loading values fall between the maximum L/D line (to the right) and the 'parasitic drag equals three times the induced drag' line (to the left) – note that the best L/D line is deep inside the region where the stall constraint is violated.

The Jetstream design points are shown here merely to illustrate the relevance of such constraint analyses to real designs – the preliminary constraint calculations we used here are based on approximations, which differ from the final aircraft in a number of ways. In this case, for instance, our simple analysis did not take into account the effect of the engines, which immerse parts of the wing and the high-lift system in their prop-wash and generate drag when not operational. In the interest of simplicity and clarity we also neglected a number of constraints (landing distances, balanced field length, turn rates, etc.); nor did we consider multidisciplinary effects – for example, the often competing demands of structural and aerodynamic design. Such details can be incorporated into a more accurate performance model, with which the constraint analysis can be revisited. In a real design exercise, multiple such iterations may be necessary before the external shape (outer mould line) of the aircraft is frozen.

In any case, to reiterate the central message of this section, the final stage in creating a preliminary design geometry is to scale the nondimensional sketch, and such scaling must be conducted on the basis of performance requirements and other design objectives. In a computational engineering framework this would typically take the shape of an optimization framework. Such frameworks could simply pose the problem as a multi-objective, multidisciplinary, constrained optimization problem.

At the highest level of the hierarchy we discussed in Section 2.7.1 this could take the form of *minimizing life cycle cost subject to a set of performance constraints*. In the absence of a sophisticated cost model, surrogate metrics (such as weight) may be used to guide the optimizer. An automated or semi-automated optimization study of this sort does not necessarily need a graphical representation of the constraints; nor does it have to be conducted in thrust-to-weight ratio versus wing loading space. Here, we chose this mode of presentation to illustrate the way in which performance requirements drive the scaling of a geometry.

Perhaps the best preliminary design geometry scaling technique is a combination of the two approaches: a black-box optimization routine to obtain the best compromise (or non-dominated compromises in a multi-objective, Pareto-type search framework) and a graphical analysis to help *understand* the design and its sensitivities with respect to alterations in the performance requirements. Of course, these requirements are often immovable (such as those resulting from certification criteria), but often they are not, and such studies might help convey to the customer the cost of, say, being able to operate from a shorter runway or further reducing the stall speed.

9.5 Indirect Wing Geometry Scaling

We are almost at the point of closing the wing geometry definition loop. We have defined a framework, whereby spanwise-varying parameters of the wing can be defined as functions of a nondimensional wing leading edge bound-coordinate axis. Thus, we have a wing of leading edge length one with the desired and potentially fully parameterized (even in terms of functional form) spanwise variations of dihedral angle, sweep angle, twist angle and aerofoil shape. All it needs now is the appropriate *scale*.

This is most easily done (from an implementation point of view) by *direct* definition of the two scale parameters we introduced in Chapter 8: `ScaleFactor` and `ChordFactor` (the key illustration here is Figure 8.18).

What if, however, our conceptual sizing process does not yield these directly? In fact, we only need to go as far as the commuter airliner case study we have just discussed for an example: the result of the constraint analysis process shown there was the projected wing area, not the scale factors. And this is quite common in aircraft design practice: the sizing process may yield a target area and a desired aspect ratio or perhaps an aspect ratio and a wetted area. Certain combinations of these may define the scale unequivocally, in which case this is what we would prefer to do, instead of just assigning `ScaleFactor` and `ChordFactor` values by trial and error.

The solution, therefore, is *indirect scaling*; that is, specifying the sorts of variables aeronautical engineers deal with more naturally than with scale factors: area, aspect ratio, span, and so on. What happens under the hood of such a geometry model is then an automated, systematic search process for finding the values of `ScaleFactor` and `ChordFactor` that give us a wing with the specified engineering parameters – an optimization problem, the variables of which are `ScaleFactor` and `ChordFactor` and the objective function is some measure of how close the parameters of the resulting wing are to the target parameters.

There is more than one way of implementing this – the code supplied with this book (e.g. the `LiftingSurface` class we have discussed at length in the last two chapters) uses a weighted squares approach. More specifically, the objective function is a linear combination of square-error-type deviations between the following parameters associated with the current wing and their target values:

- projected area
- wetted area
- span
- aspect ratio and
- root chord length.

On its own, none of the above will specify the scaling of a wing (with predefined spanwise parameter variations described by the user-defined functions) unequivocally, but any *pair* of them will. The advantage of the weighted formulation is that one may overspecify a scaling; that is, target values may be specified for more than two of the above list and the model will attempt to find the 'best-fit' wing in a weighted-least-squares sense, where the weightings may define our biases expressing the relative importance of meeting each target.[3] Of course, there has to be a good engineering reason for setting multiple targets – *entia non sunt multiplicanda since necessitate.*

---

[3] Implementation note: our OpenNURBS/Rhino-Python implementation of this method uses the `scipy` optimization routine `fmin`; at the time of writing this is not available in the Rhino environment under Mac OS X. Readers wishing to use the code should substitute another local search routine.

10

Design Sensitivities

In section 2.7.3 we briefly introduced local optimizers, which rely on gradients of the objective and/or constraint function with respect to the design variables: *design sensitivities*. These methods find their niche in problems that have just one optimum (are unimodal) and where the objective/constraint functions are quick to analyse. Unfortunately, such problems are rarely encountered and, when they are, do not remain problems for long. Local methods should also be applied when a problem has so many variables, and function evaluations take such a long time, that global search is impracticable. These situations require the efficient calculation of accurate design sensitivities for the local search algorithm to exploit. Our examples in this chapter are, in fact, of the former quick and simple type, but allow us to illustrate methods that might be applied to more difficult problems.

10.1   Analytical and Finite-Difference Sensitivities

There are a number of ways in which we might calculate derivatives with respect to the geometry. If the geometry and analysis are simple, one might be able to differentiate the whole lot analytically by hand. Clearly, this is unlikely to be achieved for CFD simulation of 3D geometry, but simple studies may be tackled this way. As an example, let us say that lift is our figure of merit and can be calculated as

$$L = qSC_L, \tag{10.1}$$

where the $S = x_0^2$ (an area calculated from one geometry variable) and $C_L = 2\pi x_1$ (the lift coefficient approximated from the angle of attack x_1). The derivatives are

$$\frac{\partial L}{\partial x_0} = 4\pi q x_0 x_1 \tag{10.2}$$

Aircraft Aerodynamic Design: Geometry and Optimization, First Edition. András Sóbester and Alexander I J Forrester.
© 2015 John Wiley & Sons, Ltd. Published 2015 by John Wiley & Sons, Ltd.

and

$$\frac{\partial L}{\partial x_1} = 2\pi q x_0^2. \tag{10.3}$$

The cost of calculating the derivatives scales with the number of variables.

If the calculation of drag is not so trivial or we do not have access to the computer code, an approximation to the derivatives can be calculated by finite differencing – for example, a one-sided forward difference

$$\frac{\partial L}{\partial x_i} = \frac{L(x_i + h) - L(x_i)}{h} + \mathcal{O}(h) \tag{10.4}$$

or a two-sided central difference

$$\frac{\partial L}{\partial x_i} = \frac{L(x_i + h) - L(x_i - h)}{2h} + \mathcal{O}(h^2), \tag{10.5}$$

where h is a small scalar and $\mathcal{O}(.)$ is the truncation error. The cost of calculating the derivatives is $n + 1$ lift calculations for one-sided differencing and $2n + 1$ lift calculations for central differencing. Subtractive cancellation limits the size of h (usually $h = \sqrt{\mathrm{eps}(x)}$, where $\mathrm{eps}(x)$ is the positive distance from $\|x\|$ to the next larger in magnitude floating-point number of the same precision as x). Figure 10.1 shows how the choice of x affects the error in the one-sided

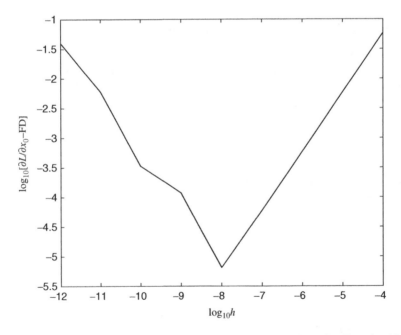

Figure 10.1 One-sided difference error when approximating $\partial L \partial x_0$ using Equation 10.4.

difference approximation of $\partial L/\partial x_0$ for our example. Depending on the function, the error will behave differently; for example, greater accuracy will result in choosing a large h for approximating $\partial L\partial x_1$, as the effect on L is linear. For solutions based on equations discretized over a mesh (e.g. CFD), a much larger h needs to be chosen even to approximate the correct sign of the gradients.

With access to the computer code, but without the ability to differentiate the formulae, the *complex step approximation* can be used to give more precise gradients:

$$\frac{\partial L}{\partial x_i} = \frac{\mathcal{I}[L(x_i + ih)]}{h} + \mathcal{O}(h^2). \tag{10.6}$$

To implement this method, we need to be able to define the design variables x_i as complex numbers and perform the calculations in the code with complex arithmetic, which is usually quite straightforward. The step h can be smaller when using the complex step approximation, and so more accurate gradients can be obtained, albeit at the extra cost of complex arithmetic.

With design optimization in mind, there are clear disadvantages with the above methods: analytical calculation is tedious and most likely infeasible, while finite differencing and complex step approximations are expensive. The simplest method, finite differencing, unfortunately has the biggest drawback: its inaccuracy. An alternative, which is growing in popularity, is *algorithmic differentiation* (AD).

10.2 Algorithmic Differentiation

Given access to the computer code, this method can yield exact derivatives and comes in two basic flavours: forward and reverse.

10.2.1 Forward Propagation of Tangents

This is essentially the incremental application of the chain rule to computer code. Table 10.1 shows the procedure for the simple lift calculation of Equation 10.1.[1] At each step in the procedure there is a value v_i and its tangent $\dot{v}_i$. We start with the geometry variables and their tangents. In this example, the tangent of the first variable is set to zero and that of the second to one; that is, we are going to calculate the partial derivative with respect to the second variable. We then proceed to apply the chain rule to each mathematical operation, which is somewhat trivial in this case, to arrive at the function value and its derivative with respect to x_1. Note that, although we have not derived an analytical expression for the derivative, the numerical derivative we have calculated is exact (up to machine precision – just as it would be if evaluating an analytical derivative). A Python implementation of Table 10.1 is given in Listing 10.1. The code is rather long-winded and could be more succinct, but serves as an illustration.

---

[1] This procedure is for a single assignment code. Dealing with multiple overwrites of memory locations is covered by Griewank (2000).

Table 10.1   Forward propagation of tangents for Equation 10.1.

$$v_0 = x_0$$
$$v_1 = x_1$$
$$\dot{v}_0 = 0$$
$$\dot{v}_1 = 1$$

$$v_2 = v_0 * v_0$$
$$\dot{v}_2 = \dot{v}_0 v_0 + v_0 \dot{v}_0$$
$$v_3 = v_1 * v_2$$
$$\dot{v}_3 = \dot{v}_2 v_1 + v_2 \dot{v}_1$$

$$y = 2\pi q v_3$$
$$\dot{y} = 2\pi q \dot{v}_3$$

At present this method does not look a lot better than differentiating the function by hand. For this example the analytical solution is indeed preferable. For more complex expressions, the analytical partial derivatives would become unwieldy, whereas the incremental numerical evaluation in AD simplifies coding. The key benefit of AD is that the chain rule can be applied automatically; indeed, AD is also known as *automatic differentiation*. There are an increasing

Listing 10.1   A Python implementation of the algorithm in Table 10.1.

```python
def lift_fwd(x,xdot):
    # initialize arrays
    v = np.zeros(5)
    vdot = np.zeros(5)
    # define aerodynamic 'constants'
    rho = 1.2
    V = 12.5
    q = 0.5 * rho * V ** 2
    # claculate constant term
    const_mult = 2 * pi * q
    # set variables at start of procedure
    v[0] = x[0]; v[1] = x[1]
    # set tangents at start of procedure
    vdot[0] = xdot[0]; vdot[1] = xdot[1]
    # step by step calculation of function and
    # its tangent
    v[2] = v[0] * v[0]
    vdot[2] = vdot[0] * v[0] + v[0] * vdot[0]
    v[3] = v[2] * v[1]
    vdot[3] = vdot[2] * v[1] + v[2] * vdot[1]
    v[4] = const_mult * v[3]
    vdot[4] = const_mult * vdot[3]
    # final function and tangent values
    y = v[4]
    ydot = vdot[4]
    return y, ydot
```

Listing 10.2   A Python function to solve Equation 10.1.

```
1  def lift(x):
2      rho = 1.2
3      V = 12.5
4      q = 0.5 * rho * V ** 2
5      y= x[0] ** 2 * x[1] * q * 2* pi
6      return y
```

number of AD tools; for example, MAD (Forth, 2006) for MATLAB® and AlgoPy for Python (Walter and Lehmann, 2013). Given the function to calculate lift in Listing 10.2, it is trivial to compute the Jacobian in Listing 10.3 using AlgoPy to yield the output:

```
jacobian =  [   589.04862255   14726.2155637 ]
```

for $x_0 = 5$ and $x_1 = 0.1$.

Runtime for forward-mode AD rises linearly with the number of directional derivatives calculated and is comparable to finite differencing in terms of speed and memory consumption.

10.2.2   Reverse Mode

Forward mode is efficient at calculating the derivatives of a large number of dependent variables (objective/constraint functions) with respect to a small number of independent variables (design variables); for example, the derivatives of the surface pressure coefficient at all mesh points with respect to a handful of geometry variables. However, we often have a large number of independent variables and only a handful of dependent variables; for example, lift, drag, cost and weight. Reverse mode is more efficient in such cases, with the cost of evaluating derivatives scaling with the number of dependent, rather than independent, variables.

Griewank (2000) is the definitive work on this subject. Here, we provide an introduction by working through our lift equation example to show the steps required for reverse differentiation. We start with a *forward sweep* to calculate intermediate values v_i and elemental functions φ_i which are stored to be used in the *reverse sweep* to follow. This forward sweep is shown for the lift equation in the first part of Table 10.2. In the reverse sweep we calculate adjoint intermediate variables $\bar{v}_j$ as

$$\bar{v}_j = \sum_{i>j} \bar{v}_i \frac{\partial \varphi_i}{\partial v_j}. \tag{10.7}$$

Listing 10.3   Forward-mode AlgoPy differentiation of Equation 10.1 (calculated by lift() in Listing 10.2.1).

```
1  from algopy import UTPM, exp
2  x = UTPM.init_jacobian([5,0.1])
3  y = lift(x)
4  algopy_jacobian = UTPM.extract_jacobian(y)
5  print 'jacobian = ',algopy_jacobian
```

Table 10.2 Incremental adjoint recursion of a single assignment code for Equation 10.1.

				CGraph[a]
Forward sweep				
v_0		$=$	x_0	`1 getitem`
v_1		$=$	x_1	`3 getitem`
v_2	$= \varphi_2$	$=$	$v_0 * v_0$	`2 pow`
v_3	$= \varphi_3$	$=$	$v_1 * v_2$	`4 mul`
y	$= \varphi_4$	$=$	$2\pi q v_3$	`6,8,10 mul`
Reverse sweep				
$\bar{v}_4$		$=$	$\bar{y}$	
$\bar{v}_3$	$+= \bar{v}_4 \frac{\partial \varphi_4}{\partial v_3}$	$=$	$\bar{v}_4 * 2\pi q$	
$\bar{v}_2$	$+= \bar{v}_3 \frac{\partial \varphi_3}{\partial v_2}$	$=$	$\bar{v}_3 * v_1$	
$\bar{v}_1$	$+= \bar{v}_3 \frac{\partial \varphi_3}{\partial v_1}$	$=$	$\bar{v}_3 * v_2$	
$\bar{v}_0$	$+= \bar{v}_2 \frac{\partial \varphi_2}{\partial v_0}$	$=$	$\bar{v}_2 * 2 * v_0$	
$\bar{x}_0$		$=$	$\bar{v}_0$	
$\bar{x}_1$		$=$	$\bar{v}_1$	

[a]Steps in the AlgoPy computational graph shown in Figure 10.2.

The notation $i \succ j$ indicates that the operation is performed for all elemental functions φ_i which are dependent on intermediate variable v_j. The first adjoint intermediate variable is initialized at the function value at which the derivatives are to be calculated; for example, the lift of the current design: $\bar{v}_4 = \bar{y}$.

The second part of Table 10.2 shows the reverse sweep of the lift equation example, yielding, in the last two rows, the derivatives in the form of the adjoints $\bar{x}_0$ and $\bar{x}_1$. The steps in Table 10.2 are implemented in Python in Listing 10.4 and can be performed automatically with AlgoPy using the code in Listing 10.5. The elemental functions are recorded in a *computational graph*, implemented in `algopy.CGraph`. Based on this graph, AlgoPy performs the forward and reverse sweeps in Table 10.2 automatically. The last two lines of Listing 10.5 print out this trace of the code and saves a graphical representation, which is shown in Figure 10.2.

'Automatic' AD is a developing area, and in the remaining examples of this chapter we will use AlgoPy in the more simple-to-implement forward mode. We expect automatic reverse mode to become faster and more easily applicable to wider ranging problems in the coming years.

10.3   Example: Differentiating an Aerofoil from Control Points to Lift Coefficient

The Python function `panel(x,y,alpha,Re)` calculates the lift coefficient c_l and pressure coefficient c_p over the surface of an aerofoil, defined by surface coordinates in an array x, y with the coordinates starting at the trailing edge, running along the lower surface to the leading edge, and back along the upper surface to the trailing edge.

Listing 10.4   A Python implementation of the algorithm in Table 10.2.

```
 1  def lift_rev(x,ybar):
 2        # initialize arrays
 3        v = np.zeros([5, 1])
 4        vbar = np.zeros([5, 1])
 5        xbar = np.zeros([2, 1])
 6        # define constants
 7        rho = 1.2
 8        V = 12.5
 9        # claculate constant term
10        const_mult = 2 * pi * 0.5 * rho * V ** 2
11        # set variables at start of forward sweep
12        v[0] = x[0]; v[1]=x[1]
13        v[2] = v[0] * v[0]
14        v[3] = v[2] * v[1]
15        v[4] = const_mult * v[3]
16        y = v[4]
17        # reverse sweep
18        vbar[4] = ybar
19        vbar[3] = vbar[4] * const_mult
20        vbar[2] = vbar[3] * v[1]
21        vbar[1] = vbar[3] * v[2]
22        vbar[0] = vbar[2] * 2 * v[0]
23        # assign gradients
24        xbar[0] = vbar[0]
25        xbar[1] = vbar[1]
26        return y, xbar
```

Listing 10.5   Reverse-mode AlgoPy differentiation of Equation 10.1 (calculated by `lift()` in Listing 10.2).

```
 1  # build computational graph
 2  cg = algopy.CGraph()
 3  x = algopy.Function([5, 0.1])
 4  y = lift(x)
 5  cg.trace_off()
 6  cg.independentFunctionList = [x]
 7  cg.dependentFunctionList = [y]
 8  # use computational graph for automatic AD
 9  print 'gradient =', cg.gradient([5,0.1])
10  # print and save computational graph
11  print cg
12  cg.plot('lift_tracer_cgraph.png')
```

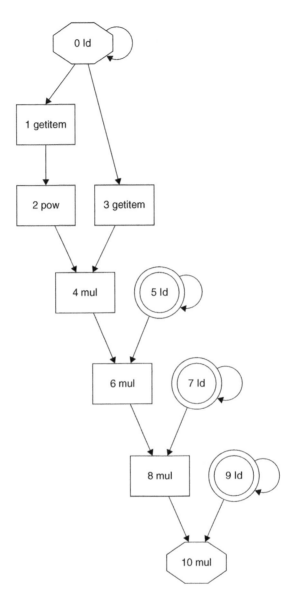

Figure 10.2   AlgoPy's computational graph of the forward part of Listing 10.2.

Listing 10.6 shows how we can use AlgoPy to differentiate `panel()` with respect to the aerofoil surface coordinates. The aerofoil surface points are defined by a Bézier spline (see Section 3.2), but could be any type of aerofoil. The aerofoil and control points are shown in Figure 10.3.

After running Listing 10.6, we can plot the derivatives, as shown in Figure 10.4. In fact, the derivatives with respect to the last five trailing edge points have not been plotted, as here numerical instabilities cause the derivative to fluctuate even more wildly.

Listing 10.6 Differentiating `panel(x,alpha,Re)` with respect to aerofoil surface coordinates.

```
 1  import numpy, algopy
 2  # using a Bezier spline aerofoil definition
 3  # define control points
 4  a=numpy.array([[1.0, 0.0],[0.75, 0.01],[0, -0.03],\
 5    [0.0, 0.0],[0.0, 0.0],[0, 0.03],[0.5, 0.05],[1.0, 0.0]])
 6  # calculate Bezier spline surface coordinates
 7  b=beziersplineaerofoil(a)
 8  # initialise Jacobian with surface coordinates
 9  x=algopy.UTPM.init_jacobian(b)
10  # define aerofoil conditions
11  Re=1e6
12  nu=1.461e-5
13  alpha=6
14  # call panel code
15  cl,cp=panel(x,alpha,Re,nu)
```

The sharp changes in the derivatives with respect to points closer to the trailing edge are in line with what we would expect with a panel solver; there is no separation, and so a cusp at the trailing edge will have the greatest effect on the lift. An optimizer would use these derivatives to exploit the lack of separation prediction in the potential flow-based solver and produce a poor aerofoil design. This is a simple example of how a parameterization and solver combination capable of producing poor designs may be exploited by an optimizer to, indeed,

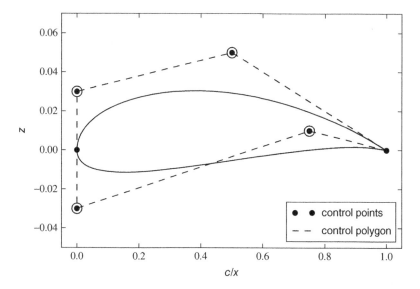

Figure 10.3 The Bézier spline aerofoil for which the derivatives of c_l with respect to its surface are to be obtained. The circled control points are those for which the z-coordinate will be varied via an inverse design process in Section 10.4.

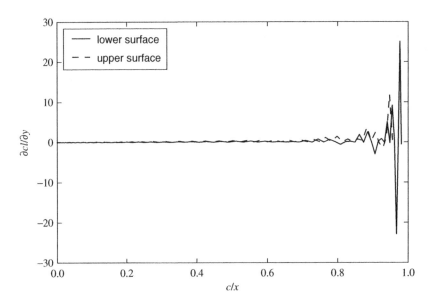

Figure 10.4 Derivatives of `panel(x,alpha,Re)` with respect to aerofoil surface coordinates.

produce poor designs. Recalling the discussion in Section 2.3, surface coordinates are a (very) flexible parameterization, but it is neither concise nor robust. These, and the obvious noise in the derivatives, are reasons why surface points are not usually a good description of aerofoils for design purposes.

At this stage we should note that there is another viewpoint here. Jameson (1999) advises to 'think in terms of free surfaces and calculus of variations' and takes the view that 'the true optimum shape belongs to an infinitely dimensional space of design parameters'. He has derived, by hand, the adjoint partial differential equations for transonic flow modelled by potential, Euler and Navier–Stokes equations. However, such an approach is difficult to implement, and the movement of surface points is not intuitive from a designer's perspective. As well as differentiating the flow solver, it is important to be able to differentiate the geometry definition to obtain derivatives with respect to more 'usable' design parameters. The better integration of the design process from CAD through to analysis does, however, fit in with the Jameson (1999) view of the future, and the use of reverse-mode AD offers the prospect of adjoint codes, with sensitivities propagated from design parameter through to a range of multidisciplinary analyses.

Derivatives of the aerofoil surface with respect to the Bézier spline control points in Figure 10.3 are straightforward to calculate with AlgoPy by initializing the Jacobian with the control points (as shown in Listing 10.7) and the resulting $\partial B/\partial a_i$ values are shown in Figure 10.5.

Differentiating a Bézier curve (Equation 3.12) with respect to its control points yields the Bernstein polynomials and, comparing Figure 10.5 with Figure 3.7, we see that we have indeed calculated these, but they are plotted versus Cartesian x/c rather than curve parameter u, and so there is a difference between the upper and lower surface derivatives.

Listing 10.7 Differentiating `panel(x,alpha,Re)` with respect to Bézier spline control points.

```
1  import numpy, algopy
2  # initialise Jacobian with control point locations
3  a=UTPM.init_jacobian([[1.0, 0.0],[0.75, 0.01],[0, -0.03],\
4       [0.0, 0.0],[0.0, 0.0],[0, 0.03],[0.5, 0.05],[1.0, 0.0]])
5  # calculate Bezier spline surface coordinates
6  b=beziersplineaerofoil(a)
7  jacobian = UTPM.extract_jacobian(b)
```

As a brief aside, we can differentiate other aerofoil definitions to yield similarly intuitive results. Figures 10.6 and 10.7 show an aerofoil defined by upper and lower Ferguson splines (see Section 3.3) and the derivative of its surface with respect to the Ferguson spline parameters. The derivatives are the Ferguson spline basis functions (see Figure 3.12), but are plotted in Figure 10.7 with respect to x/c rather than the curve parameter u. Using the definition in Section 7.2, we obtain the more intuitive derivatives with respect to tangent 'tensions' and angles shown in Figure 10.8.

As another example, Figure 10.9 shows the derivatives of the z-coordinate of a NACA 4412 aerofoil (see Section 6.1) with respect to the four-digit definition, which are the camber distribution ($\partial z/\partial z_{cam}^{max}$), the thickness distribution ($\partial z/\partial t_{max}$) and the slightly less obvious effect of the maximum camber location ($\partial z/\partial x_{mc}$). For this, the Bézier and Ferguson spline, the derivatives are straightforward, but serve as an example of the power of AD as a tool for propagating derivatives.

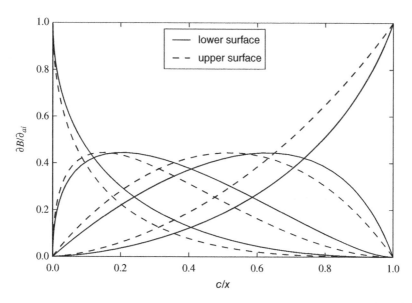

Figure 10.5 Derivatives of the Bézier spline aerofoil surface with respect to the control points (compare with the Bernstein polynomials in Figure 3.7).

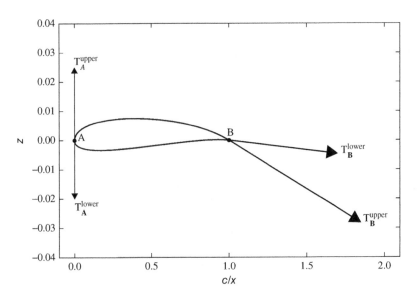

Figure 10.6 A Ferguson spline aerofoil defined by $\mathbf{A}^{\mathrm{upper,lower}} = 0.0, 0.0$, $\mathbf{B}^{\mathrm{upper,lower}} = 1.0, 0.0$, $\mathbf{T}_{\mathbf{A}}^{\mathrm{lower}} = 0.0, -0.025$, $\mathbf{T}_{\mathbf{A}}^{\mathrm{upper}} = 0.0, 0.03$, $\mathbf{T}_{\mathbf{B}}^{\mathrm{lower}} = 0.75, 0.0$, $\mathbf{T}_{\mathbf{B}}^{\mathrm{upper}} = 0.9, -0.03$.

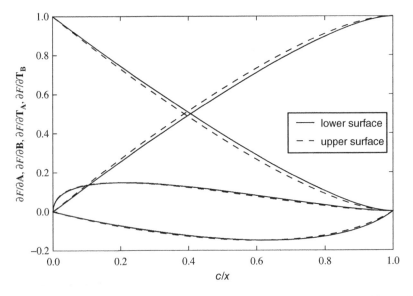

Figure 10.7 Derivatives of the Ferguson spline aerofoil surface with respect to the leading and trailing edge points (**A** and **B**), and leading and trailing edge tangents ($\mathbf{T}_{\mathbf{A}}$ and $\mathbf{T}_{\mathbf{B}}$) (compare with the basis functions in Figure 3.12).

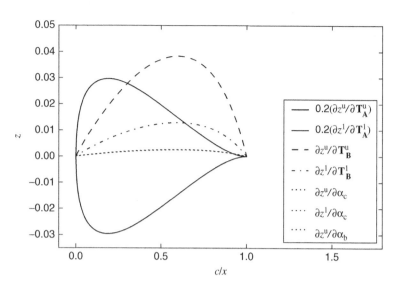

Figure 10.8 Derivatives of the Ferguson spline aerofoil surface with respect to the definition in Figure 7.2, showing the intuitive nature of this parameterization (i.e. the variables have clear associations with the shape of the aerofoil). Note that although some parameters have the same effect on the shape, they will, naturally, have different effects on the flow.

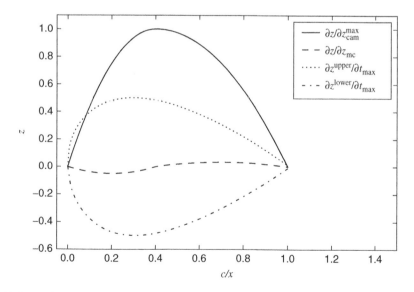

Figure 10.9 Derivatives of a NACA 4412 aerofoil surface with respect to the four-digit definition; that is, z_{cam}^{max}, x_{mc} and t_{max}.

Back to our Bézier spline example, having differentiated the geometry in Listing 10.7, the derivatives of c_1 with respect to the control points are found by calling cl,cp=panel(p,alpha,Re,nu) and extracting the Jacobian to yield:[2]

$$\frac{\partial c_1}{\partial \mathbf{a}} = \begin{pmatrix} -1.71e+02 & 9.48e+02 \\ -1.23e-01 & 1.38e-01 \\ 3.03e-02 & 1.29e+00 \\ -5.61e-01 & 1.61e+00 \\ -1.62e-01 & 3.52e-01 \\ 2.09e-02 & 4.45e-01 \\ 4.25e-01 & 4.78e+00 \\ 1.71e+02 & -9.57e+02 \end{pmatrix}. \tag{10.8}$$

The lift coefficient is most sensitive to the trailing edge points ($\frac{\partial c_1}{\partial a_0} \neq \frac{\partial c_1}{\partial a_7}$) as the panel code control points are near to, but not coincident with, the trailing edge).

We can go further and calculate the derivatives of the drag with respect to the control points by calling cd=boundarylayer(v,ds,xc,yc,theta,Re,alpha), having first obtained the necessary inputs by requesting further outputs from the panel code: cl,cp, v,ds,xc,yc,theta=panel(p,alpha,Re,nu). The viscous drag is calculated in boundarylayer() by estimating the boundary-layer thickness using the 'e^9' method (see Drela (2014) and also the useful thesis by Wauquiez (2000)). The derivatives calculated through our AlgoPy implementation are

$$\frac{\partial c_d}{\partial \mathbf{a}} = \begin{pmatrix} -1.33e+01 & -7.54e+00 \\ 1.92e-03 & 7.84e-03 \\ 1.02e-03 & -6.21e-04 \\ 1.66e-04 & -5.97e-02 \\ -3.98e-03 & 5.81e-02 \\ 1.45e-03 & 3.23e-03 \\ -1.61e-03 & -8.18e-03 \\ 1.33e+01 & 7.54e+00 \end{pmatrix}. \tag{10.9}$$

10.4   Example Inverse Design

In the call to panel() above, we have, among other quantities, calculated the derivatives of the pressure coefficient, which may be of interest to the designer, particularly in an inverse design process (see also Section 2.7.4). The pressure coefficient along the surface of the aerofoil is shown in Figure 10.10.[3] Also shown is a target pressure profile that we can obtain via a

---

[2] The Jacobian is a 1×16 vector ($\mathcal{J} \in \mathbb{R}^{m \times n}$ for $F : \mathbb{R}^n \to \mathbb{R}^m$), but we have shown the derivatives in an 8×2 (control points $\times$ x, z-coordinate format).

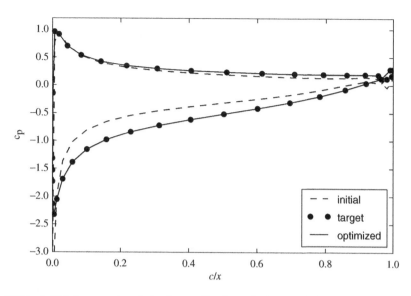

Figure 10.10   Initial, target and optimized c_p profiles from the inverse design process in Listing 10.9.

decoupled inverse design process[4] by minimizing the sum of the squares of the difference between the target and current pressure profiles:

$$\min_{\mathbf{a}} \sum \left(\mathbf{c}_p^T - \mathbf{c}_p(\mathbf{a})^K \right)^2 , \qquad (10.10)$$

where $\mathbf{a}$ is a vector of control points to be modified in an attempt to obtain min $= 0$ and $\mathbf{c}_p^T$ and $\mathbf{c}_p(\mathbf{a})^K$ are vectors containing the target and current pressure profiles. We will solve this problem by varying the z-locations of control points above and below the leading edge at $x = 0$, the lower surface point at $x = 0.75$ and the upper surface point at $x = 0.5$ (these points are circled in Figure 10.3). Although decoupled, with the derivatives of Equation 10.10 with respect to the control points, the inverse design problem can be solved with only a few steps of the SciPy implementation of the BFGS algorithm (Zhu *et al.*, 1997).

Listing 10.8 is a function `invdes_der()`, that modifies a `basefoil` according to the z-locations defined in `x`, calculates its pressure profile with a call to `panel()` and then

---

[3] The c_p should go to one at the leading and trailing edge stagnation points; however, the first and last (rearmost lower and upper) control points are not at the trailing edge (they are at the centre of the panels at the trailing edge), and so we cannot expect $c_p = 1$, particularly when a small number of panels are used (here 41).

[4] A *decoupled* process refers to our formulating the inverse design problem in Equation 10.10 by comparing pressure profiles calculated by independent runs of a panel code; that is, there is no mathematical link between $\mathbf{c}_p^T$ and the geometry. A *coupled* inverse design method (e.g. available in XFOIL – see Chapter 11) calculates the geometry directly via the solution of a boundary-value problem. A coupled approach is computationally more efficient, but a decoupled method can be employed with any existing flow solver – here the Python function `panel()`.

Listing 10.8 A Python function using AlgoPy to calculate the derivatives of Equation 10.10 with respect to the z-locations of the control points circled in Figure 10.3.

```
 1  def invdes_der(x,targetcp,basefoil,Re,nu,alpha):
 2      b=basefoil
 3      for i in range(0,2):
 4          b[i+1,1]=x[i]
 5          b[i+5,1]=x[i+2]
 6      a=UTPM.init_jacobian(b)
 7      p=beziersplineaerofoil(a)
 8      [cl,cp,v,ds,xc,yc,theta]=panel(p,alpha,Re,nu)
 9      diff=sum((cp-targetcp)**2)
10      jacobian = UTPM.extract_jacobian(diff)
11      d=jacobian[ix_([3,5,11,13])]
12      return d
```

computes Equation 10.10 with this profile and `targetcp`. The control points of the aerofoil are defined as AlgoPy's `UTPM.init_jacobian`, and the derivatives with respect to the z-locations are returned by the function. In the second to last line, the `ix_` command is used to extract elements `[3,5,11,13]`, which correspond to the circled control points in Figure 10.3, from the Jacobian of 16 derivatives with respect to all Bézier spline control point x, y locations).

Listing 10.9 implements a BFGS optimization of Equation 10.10, yielding the following result:

```
(array([ 0.01499992, -0.02999998,  0.06999999,  0.09999996]),
1.4489377963155702e-12,
{'warnflag': 0,
'task': 'CONVERGENCE: REL_REDUCTION_OF_F_<=_FACTR*EPSMCH',
'grad': array([  5.38809059e-07,   6.83056095e-05,
  -1.30611260e-04, -1.20647665e-05]),
'nit': 23, 'funcalls': 40})
```

Listing 10.9 Python code to solve the inverse design problem using `scipy.optimize.fmin_l_bfgs_b` (invdes), called by the optimizer, is similar to Listing 10.8, but without the AlgoPy Jacobian initialization and derivative extraction.

```
 1  basefoil=array([[1.0, 0.0],[0.75, 0.01],[0, -0.03],\
 2      [0.0, 0.0],[0.0, 0.0],[0, 0.03],[0.5, 0.05],[1.0, 0.0]])
 3  Re=1e6
 4  nu=1.461e-5
 5  alpha=5
 6  fargs=[targetcp, basefoil, Re, nu, alpha]
 7  x=array([0.01,-0.03,0.03,0.05])
 8  res=fmin_l_bfgs_b(invdes,x,fprime=invdes_der,args=fargs,disp=True)
```

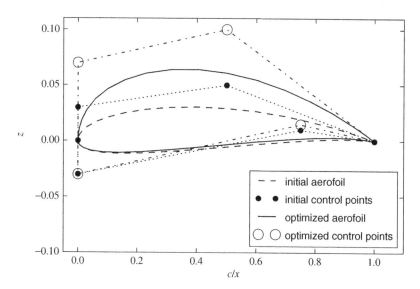

Figure 10.11   Initial and optimized aerofoils from the inverse design process in Listing 10.9.

The residual sum of square differences between the target and optimized pressure profiles is $\mathcal{O}(10^{-12})$, and the close agreement can be seen visually in Figure 10.10, which shows the initial, target and optimized pressure profiles. The initial and optimized aerofoils are shown in Figure 10.11.[5]

---

[5] We actually cheated slightly and calculated the target c_p based on the solution of an aerofoil with x=[0.015, -0.03, 0.07 0.1], which the optimizer has found via matching the pressure profiles. Unless a 'free surface' is used, and the adjoint of Equation 10.10 with respect to this surface, a true inverse design process is unlikely to achieve such a good c_p match.

11

Basic Aerofoil Analysis: A Worked Example

We have illustrated the geometrical concepts described in this book through examples wherever possible, but have yet to close the loop; that is, link these concepts into complete design processes. In this chapter and the next we endeavour to do just that, first through a simple 2D flow solver coupled with an aerofoil parameter sweep, then by taking that example further to the aero/structural design of a wing planform and sections.

Despite the growing use of whole configuration, 3D CFD, with increasingly accessible solvers, such as OpenFOAM (Jasak *et al.*, 2007), 2D analysis of aerofoils still has its place in the design process, particularly for high aspect ratio non-swept wings. XFOIL is an interactive program for the design and analysis of subsonic isolated aerofoils (Drela, 1989). It combines the speed and accuracy of high-order panel methods with a fully coupled viscous boundary-layer method (Drela and Giles, 1987). The software is freely available to download from Drela's website (Drela and Youngren, 2011). Here, we will consider XFOIL as a black-box analysis method and use it to demonstrate geometry manipulation and optimization (we will *use* it but not delve into its inner workings). Simpler, nonvalidated Python and MATLAB® codes (`panel()` and `boundarylayer()`) are included in the toolkit accompanying this book for those looking for closer integration between geometry and solver (e.g. through the use of algorithmic differentiation – see Chapter 10).

XFOIL can be run interactively at the command prompt using a set of commands detailed in the user guide (Drela and Youngren, 2011), where a sample session can also be found. Listing 11.1 is a sample session that runs a viscous analysis of an aerofoil defined in `aerofoil.dat`. The inputs work, line by line, as follows:

- load aerofoil coordinates from the `aerofoil.dat` file,
- name this aerofoil `CubicBez`,
- smooth the aerofoil using the `panel` command,
- enter the analysis routine using the `oper` command,
- define a viscous analysis at a Reynolds number of 1.4×10^6,
- set the Mach number to 0.1,

Aircraft Aerodynamic Design: Geometry and Optimization, First Edition. András Sóbester and Alexander I J Forrester.
© 2015 John Wiley & Sons, Ltd. Published 2015 by John Wiley & Sons, Ltd.

Listing 11.1   Sample XFOIL input file.

```
 1  load C:\Users\you\xfoil\aerofoil.dat
 2  CubicBez
 3  panel
 4  oper
 5  visc 1397535
 6  M 0.1
 7  type 1
 8  pacc
 9  C:\Users\you\xfoil\polar.dat
10
11  iter
12  5000
13  cl 1.2
14
15
16  quit
```

- `type 1` means that the Mach number, Reynolds number, aerofoil chord and velocity will be fixed and the lift will vary (other types of analysis are described in the XFOIL user guide, section 5.7),
- store data in a file using the `pacc` command (short for polar accumulate),
- set `polar.dat` as the file in which to store data,
- return to ignore a prompt at this point,
- iterate the analysis 5000 times,
- set a target lift coefficient for the analysis of 1.2 (i.e. the angle of attack will be varied automatically to obtain this lift coefficient),
- return to come out of `oper` menu, and
- quit XFOIL.

Instead of entering these commands directly into the XFOIL prompt, the program can be run in batch mode by saving the commands in Listing 11.1 in a file called, say, `commands.in` and typing `xfoil_path\xfoil < file_path\commands.in` at the DOS prompt (replace '\' with '/' for LINUX/Mac operating systems). Depending on the operating system, XFOIL requires a slightly different input file format (Listing 11.1 is for Windows) – if in doubt, run through the lines of `commands.in` interactively in the command prompt first.

Listing 11.2 is an example of the `polar.dat` output file. The format of this file is always the same, which is useful to know if you are going to interrogate it for data. If the output file entered in XFOIL already exists, the `pacc` command in Listing 11.1 will fail. It is therefore best to use a new output file name or delete the existing output file before XFOIL is run again.

11.1   Creating the `.dat` and `.in` files using Python

The `aerofoil.dat` file contains the x, z coordinates of the aerofoil in two columns (x and z), with the coordinates starting at the trailing edge, running along the lower surface to the

Listing 11.2   Sample XFOIL output file.

```
 1
 2          XFOIL          Version 6.97
 3
 4    Calculated polar for: CubicBez
 5
 6    1 1 Reynolds number fixed           Mach number fixed
 7
 8    xtrf =    1.000 (top)         1.000 (bottom)
 9    Mach =    0.100      Re =     1.398 e 6      Ncrit =    9.000
10
11       alpha     CL        CD        CDp       CM       Top_Xtr  Bot_Xtr
12       ------  --------  -------   -------   -------   -------  -------
13       3.699   1.2000   0.00745   0.00290  -0.1629   0.4208   1.0000
```

leading edge and back along the upper surface to the trailing edge. We can write these aerofoil coordinates to file using the `savetxt` function (Listing 11.3). Note that the last element of the `lower` array is omitted so that the leading edge point is not duplicated.

While with the `aerofoil.dat` file we are writing columns of numbers from which the shape of the aerofoil is not immediately apparent, the input file is 'readable' (at least to the trained eye), and we wish to preserve this in the Python code via which we write it. In Listing 11.4 the `write` function is used in a format where the structure of the resulting file is apparent and the variable names (e.g. `Re` for Reynolds number) are logical. Writing code in this way is not particularly succinct, but is often preferable from a development perspective.

11.2   Running XFOIL from Python

System calls can be made from Python after importing `os` using `os.system()`. For example, assuming `xfoil` is in `C:\apps\XFOIL\` and `commands.in` and `aerofoil.dat` are in `C:\Users\you\xfoil\`, an output similar to Listing 11.2 could be produced with the code in Listing 11.5. To avoid worrying about operating systems and entering '\' or '/', the `os.path.sep` function has been used.

Now we need to retrieve the results from `polar.dat` (Listing 11.2), which can be read using, for example, the Python `readlines()` function as shown in Listing 11.6. Here, we read all the lines of the file and then assign specific elements of the last line to `cl` and `cd`. A command to delete the `polar.dat` file is included to allow XFOIL to be run again in the same way if needed.

Listing 11.3   Line-by-line writing of aerofoil coordinates.

```
1  data_file = open(file_path + 'aerofoil.dat', 'w')
2  savetxt(file_path + 'aerofoil.dat', vstack([lower[0:-1], upper]))
3  data_file.close()
```

Listing 11.4 Writing the `commands.in` file in Python in an easily identifiable format.

```
1  command_file=open(file_path + 'commands.in','w')
2  command_file.write('load ' + file_path + 'aerofoil.dat\n\
3  CubicBez\n\
4  panel\n\
5  oper\n\
6  visc ' + str(Re) + '\n\
7  M ' + str(M) + '\n\
8  type 1\n\
9  pacc\n'\
10 + file_path + 'polar.dat\n\
11 \n\
12 iter\n 5000\n\
13 cl 1.2\n\
14 \n\
15 \n\
16 quit\n')
17 command_file.close()
```

Listing 11.5 Sample Python code to run XFOIL with `os.path.sep` command.

```
1  import os
2  sep=os.path.sep
3  file_path = 'C:' + sep + 'Users' + sep + 'you' + sep + \
4   'xfoil' + sep
5  xfoil_path =' C:' + sep + 'apps' + sep + 'XFOIL' + sep
6  run_xfoil_command = xfoil_path + 'xfoil < ' + file_path + \
7   'commands.in'
8  os.system(run_xfoil_command)
```

Listing 11.6 Sample Python code to read results from `polar.dat`.

```
1  aero_data_file = open(file_path + 'polar.dat', 'r')
2  lines = aero_data_file.readlines()
3  aero_data_file.close()
4  #delete Xfoil output file ready for next Xfoil run
5  os.system('del ' + file_path + 'polar.dat')
6  #Linux/OSX: os.system('rm -f ' + file_path + 'polar.dat')
7  alpha = float(lines[-1][2: 8])
8  cl = float(lines[-1][11: 17])
9  cd = float(lines[-1][20: 27])
```

Listing 11.7   Sample Python code to run XFOIL for a variety of Ferguson spline aerofoils with varying T_A^{upper}.

```
1  file_path='...'
2  xfoil_path='...'
3  cl = 0.8433 #target lift
4  v = 12.5 #velocity
5  L = 0.698 #chord length of aircraft wing
6  nu = 0.00001461 #kinematic viscosity
7  TA_upper=linspace(0.2, 0.6, 6) # set of T_A^upper angles
8  cd=zeros(max(shape(TA_upper))) # pre-allocate cd array
9  # iterate over T_A^upper angles
10 for i in range(0,max(shape(TA_upper))):
11     # array of inputs for Ferguson definition
12     a=array([TA_upper[i], 0.3, 1.0, 1.0, 5, 15])
13     # calculate and plot Ferguson aerfoil surface points
14     points=hermite_airfoil(a)
15     plot(points[:,0],points[:,1])
16     # run XFOIL
17     [cd[i],cl]=runxfoil(points,cl,v,L,nu,file_path,xfoil_path)
```

The Python and MATLAB® functions [cd,cl,alpha]=runxfoil(points,cl,v,L, nu,file_path,xfoil_path) 'wrap' XFOIL using the above procedure. Using this and hermite_airfoil() (see Sections 3.3 and 7.2), Listing 11.7 calculates and plots the drag coefficient of a Ferguson spline aerofoil for a range of upper leading edge tangent tensions T_A^{upper}. The flow conditions are set to that for the human-powered aircraft example in Chapter 12. The aerofoils and their drag values are shown in Figures 11.1 and 11.2. Given these outputs, we could quickly pick off the optimum aerofoil.

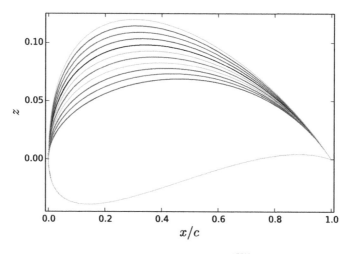

Figure 11.1   Ferguson spline aerofoils with varying T_A^{upper}, produced by Listing 11.7.

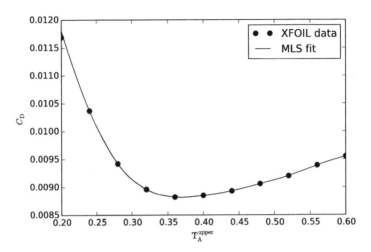

Figure 11.2 XFOIL-calculated drag coefficients for Ferguson spline aerofoils with varying T_A^{upper}, produced by Listing 11.7. The curve fit is a 'moving least squares' – for example, see Forrester and Keane (2009).

Of course, there will in fact be close interactions between all six of the Ferguson spline parameters. This six-dimensional design space could also be sampled with a 'sprinkling' of tentative aerofoil designs, and a six-dimensional curve-fit employed to guide us to the optimum (a surrogate modelling approach – e.g. see Forrester *et al.* (2008)), or an optimizer may be employed, which calls XFOIL directly.

Having covered the basics of automated calls to XFOIL, in Chapter 12 we will now indeed go on to embed this free, simple-to-use, yet powerful, tool in an optmization process.

12

Human-Powered Aircraft Wing Design: A Case Study in Aerodynamic Shape Optimization

Human-powered aircraft design is a caricature of the classic structural efficiency (thick, short wing) versus aerodynamic efficiency (thin, long wing) trade-off. The 'sport' is going through a renaissance in the UK, with the advent of the Royal Aeronautical Society's Icarus Cup, and the authors have been involved with the design of SUHPA (Southampton University Human Powered Aircraft), pictured in Figures 12.1 and 12.2. Here, we will follow a basic planform optimization (see Chapter 8) and demonstrate the use of two aerofoil parameterization strategies from Chapters 3 and 7 to optimize a wing for SUHPA for a steady, level flight speed of 12.5 m/s (world record speed) and a wing loading of 75 N/m² (based on previous aircraft data). The fluid dynamics analysis will be based on XFOIL, with which we have already showed how to analyse parametric aerofoils in Chapter 11.

The wing, of span b, is to be constructed around a spar with constant width and uniformly tapered height, with a foam core and pultruded carbon capping. The skin, along with the internal structure to carry aerodynamic loads to the spar, is hot-wire cut from styrofoam in six uniformly tapered sections. In our design, we therefore have control of the root and tip spar height (which equates to the wing thickness), the chords at the root, $b/6$ and $b/3$ stations, and the shape of a linearly varying aerofoil section between each of these stations and the wing tip.

We will make a few simplifications in our analysis of the wing.

1. The effect on the weight of the aircraft as we vary the wing design will be purely due to the quantity of pultruded carbon capping on the wing spar; that is, the foam and shear web weights per unit span will be unaffected.

Aircraft Aerodynamic Design: Geometry and Optimization, First Edition. András Sóbester and Alexander I J Forrester.
© 2015 John Wiley & Sons, Ltd. Published 2015 by John Wiley & Sons, Ltd.

Figure 12.1   SUHPA in flight during the 2013 Icarus Cup at Sywell Aerodrome (piloted by Bill Brooks; power-plant, Guy Martin; photograph by Fred To).

2. The bending stiffness of the wing is purely due to the pultruded carbon capping, and a tip deflection limit will be calculated using Euler–Bernoulli beam theory.
3. The drag is decoupled into induced drag and viscous drag calculated using a series of 2D boundary-layer approximations with XFOIL.
4. The induced drag is calculated simply from Equation 8.3: $C_D^i = \dfrac{C_L^2}{\pi e \mathrm{AR}}$.

Figure 12.2   SUHPA in flight. Note the high aspect ratio and low thickness/chord, highlighting the importance of the aero-structural trade-off (piloted by Bill Brooks; power-plant, Guy Martin; photograph by Fred To).

Table 12.1   SUHPA wing design problem set-up.

Parameter	Symbol	Units	Lower bound	Starting value	Upper bound
Flight speed	U_∞	m/s	—	12.5	—
All-up weight[a]	W	N	—	850	—
Wing loading	n	N/m^2	—	75	—
Span	b	m	16	20	24
Root thickness	t_R	mm	60	100	40
Tip thickness	t_T	mm	20	30	40
Maximum aerofoil thickness to chord[b]	$(t/c)_{max}$		—	—	0.14

[a]Not including weight of carbon pultrusion.
[b]Maximum t/c imposed as boundary-layer code performs poorly above this value.

12.1   Constraints

The only aerodynamic constraint to be imposed is that lift must equal weight at the cruise speed. The thrust-to-weight ratio is not considered, as the natural ability of the human 'powerplant' to give extra short burst of power (in the region of three times cruise power) means that sufficient climb rates can be achieved with the aircraft optimized for cruise.

A vertical wingtip deflection limit of 0.6 m at $1g$ is imposed to give sufficient dihedral for stability, yet limit aerodynamic losses. When this constraint is satisfied, at $2g$ the maximum allowable strains in the composite spar are not exceeded and so do not form part of the design process (i.e. the structural design is dominated by the stiffness requirement).

Rather than the more usual airport compatibility constraints on size, we are limited by transport costs; needing to fit the aircraft inside a standard 4 m internal capacity long-wheelbase van. With six wing sections, the upper limit on the span is therefore 24 m.

The conditions and variables are defined in Table 12.1.

12.2   Planform Design

The planform is designed following the basic procedures in Section 8.5; that is, we create a nondimensional planform definition and size this to fit our requirements. The definition is that each half-span has three non-swept sections defined by $c^{(0,1,2,3)}$ (four chords: root, two section joints, and tip). For a given span (b, one of our optimization variables) and weight estimate, the process of sizing this wing is shown in Table 12.2.

To find the unknown chord lengths, we minimize the difference between our six-section planform and the ideal elliptic planform:

$$\min_{c^{(0,1,2)}} \int_0^{b/2} c(y) - \sqrt{(1 - (2y/b)^2)}\, dy. \tag{12.1}$$

The function getchords() performs this operation.

Table 12.2 Planform sizing of SUHPA wing.

Inputs		b, W
Sizing variables (unknown)		$c^{(0,1,2)}$
L	$=$	W
S	$=$	L/n
Section length (l)	$=$	$b/6$
Inboard taper ($\lambda^{(0)}$)	$=$	$c^{(1)}/c^{(1)}$
Mid taper ($\lambda^{(1)}$)	$=$	$c^{(2)}/c^{(0)}$
Inboard area ($S^{(0)}$)	$=$	$(c^{(0)} + c^{(1)})l/2$
Mid area ($S^{(1)}$)	$=$	$(c^{(1)} + c^{(2)})l/2$
Outboard area ($S^{(2)}$)	$=$	$S/2 - S^{(0)} - S^{(1)}$
$c^{(3)}$	$=$	$(2S^{(2)}/l) - c^{(2)}$

12.3 Aerofoil Section Design

We could have created an elliptic lift distribution with a uniform chord by varying the lift coefficient according to an elliptic distribution. Varying the planform is more structurally efficient. With only six wing sections, the resulting planform from minimizing Equation 12.1 is naturally not elliptical, but now the local lift coefficient can be varied to improve the efficiency of the wing:

$$c_L^{(i)} = \frac{2L}{\rho V^2} \frac{\sqrt{(1 - (2y/b)^2)}}{c^{(i)}}. \tag{12.2}$$

The aerofoil shape is then optimized at the four spanwise locations in terms of minimizing the drag for the lift coefficient defined by Equation 12.2.

12.4 Optimization

With the planform and aerofoil design procedures defined, we can now approach the overall multidisciplinary optimization (two disciplines in fact: structural and aerodynamic). We wish to minimize drag subject to the deflection at the wingtip being less than 0.6 m. The deflection minus the constraint limit is calculated by the function suhpaconstraints(x), where x is a vector with elements $[b, t_R, t_T]$; that is, the function will return a negative value when the constraint is satisfied.

The drag is calculated using the procedures described in the previous sections, as follows:

1. The lift is found from W plus the spar and foam weight, which is simple to calculate using the pultruded carbon volume and density plus an estimate of 0.5 kg of foam per metre of span (wingweight(x)).
2. The area is found from the lift and wing loading, and the ideal, elliptic chord distribution is calculated.
3. The best fit to the ideal chord distribution possible with the six wing sections is found using getchords(x) (where x is a vector of the chords $c^{(0,1,2)}$ at the 0, $b/6$ and $b/3$ stations) and the appropriate lift coefficient for these sections is calculated.

4. Aerofoil surface coordinates for the chosen parameterization are passed to `runxfoil()`
to calculate the 2D viscous drag, which is integrated over the planform area and added to
an estimate for the induced drag.

12.4.1 NACA Four-Digit Wing

The above process is implemented for the NACA 44xx aerofoil definition in `suhpadrag()`
(where inputs are as for `suhpaconstraints(x)`). With the objective and constraint functions defined, the wing can be optimized with, for example, MATLAB's® `ga`, `fmincon` or
`scipy.optimize.fmin_l_bfgs_b`. Listing 12.1 shows the *MATLAB*® script to optimize
the NACA 44xx wing and the resulting variables are (note that, owing to the random elements
of the genetic algorithm, it is unlikely the reader will obtain identical values):

```
NACA44xxvars =

    21.8801    99.5640    38.1337



NACA44xxdrag =

    19.4062
```

Listing 12.1 Sample MATLAB® code to optimize SUHPA's wing by varying the thickness and span
using a NACA 44xx aerofoil definition.

```
 1   % set global parameters
 2   global wingParameters constants
 3   % constants
 4   constants.g=9.81; % accelratio due to gravity [ms^-2]
 5   constants.rho=1.2; % density of air [kgm^-3]
 6   constants.nu=1.461e-5; % kinematic viscosity of air [m^2s^-1]
 7   constants.V=12.5; % flight speed [ms^-1]
 8   constants.xfoil_path='...'; % file path
 9   constants.file_path='...'; % Xfoil path
10   % wing parameters
11   wingParameters.loading=75; % wing loading [Nm^-2]
12   wingParameters.e=0.85;% Oswald efficiency (estimate!)
13   wingParameters.wSpar=50; % width of spar [mm]
14   wingParameters.tCap=2.0; % thickness of carbon pulstrusion [mm]
15   wingParameters.weight=850; % weight of aircraft minus wing [N]
16   wingParameters.deflection=600; % maximum deflection at tip [mm]
17
18   % set options to follow GA search with fmincon
19   options=gaoptimset('Display','iter','HybridFnc',@fmincon);
20   % call optimizers
21   [NACA44xxvars,NACA44xxdrag]=ga(@suhpadrag,3,[],[],[],[],...
22   [16 80 20],[24 120 40],@suhpaconstraints,options)
```

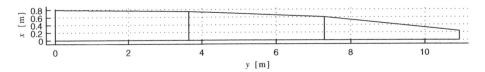

Figure 12.3 Planform of SUHPA after the three-variable NACA 44xx optimization (Listing 12.1).

which, as they should be, are close to the deflection constraint, with `wingdeflection(vars)` yielding `573.4993`; that is, within 7 mm. The planform is shown in Figure 12.3. The aerofoils at the four spanwise locations correspond to the NACA 4412.5, NACA 4410.5, NACA 4409.7 and NACA 4416.4 (with a linear variation between). These are shown in Figure 12.4, along with their pressure profiles (calculated with `panel()`).

We can, of course, vary the other two parameters in the NACA four-digit definition: z_{cam}^{max} and x_{mc}. This is accomplished with `wingParameters.foil='NACAxxxx'`, increasing the number of design variables to five, and applying appropriate upper and lower bounds: $z_{cam}^{max} \in [1, 5]$ and $x_{mc} \in [1, 5]$. The result is

```
NACAxxxxvars =

    22.5989   102.7538    38.6725     4.9996     4.1259


drag =

    18.4514
```

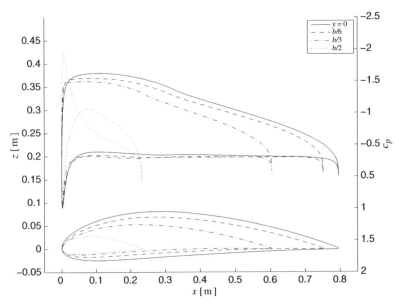

Figure 12.4 Aerofoils and corresponding pressure profiles after the three-variable NACA 44xx optimization (Listing 12.1).

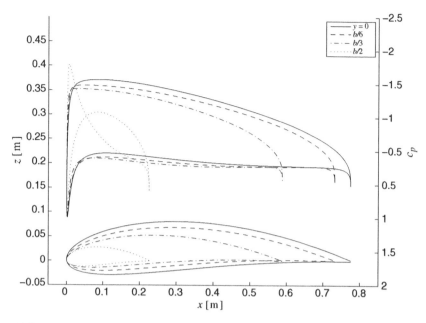

Figure 12.5 Aerofoils and corresponding pressure profiles after the five-variable NACA xxxx optimization.

The span has increased and the wing thickened, owing to the aero-structural trade-off shifting because a lower drag aerofoil design has been found. The most significant change has been the increase in x_{xc}. The aerofoils at the four spanwise locations correspond to a NACA (4.13)(5)(13.3), NACA (4.13)(5)(11.1), NACA (4.13)(5)(10.21) and NACA (4.13)(5)(17.11) and are shown in Figure 12.5. Note how moving the maximum camber location aft has created a less aggressive adverse pressure gradient on the upper surface of the aerofoils; transition is delayed, and so the drag is reduced significantly (see discussion of laminar flow in Section 5.2.5).

12.4.2   Ferguson Spline Wing

Further flexibility in the geometry, in the hope of getting closer to an optimal aerofoil design, can be added by using the Ferguson spline parameterization definition in Section 7.2. Along with b, t_R and t_T we now have six further variables T_A^{upper}, T_A^{lower}, T_B^{upper}, T_B^{lower}, α_c and α_b (shown in Figure 7.2). Without a thickness-to-chord ratio in the definition, this must be calculated and a constraint added to ensure the spar fits inside the wing. The function hermite_foil outputs t/c at $x/c = 0.25$, and this is called by suhpaconstraints() if wingParameters.foil='hermite'. Listing 12.2[1] yields the following result for the

---

[1] Note that we have set the optimizer's finite-difference step size. This is quite large at *O*0.001 to try to avoid issues with noise in the output from XFOIL corrupting the gradient calculations.

Listing 12.2   Sample MATLAB® code to optimize SUHPA's wing by varying the thickness, span and a Ferguson spline aerofoil definition.

```
 1  wingParameters.foil='hermite'; % airfoil defintion type
 2  % call GA
 3  options=gaoptimset('Display','iter');
 4  [HermitevarsGA,Hermitedrag]=ga(@suhpadrag,9,[],[],[],[],...
 5  [16 80 20 0.25 0.25 0.5 0.5 -5 10],[24 160 60 1 1 3 3 5 20],...
 6  @suhpaconstraints,options)
 7  % set finite difference step
 8  options=optimset('Display','iter','FinDiffRelStep',0.001,...
 9  'TypicalX',[16 80 20 0.25 0.25 0.5 0.5 -5 10]);
10  % call fmincon
11  [Hermitevars9,Hermitedrag9]=fmincon(@suhpadrag,HermitevarsGA,...
12  [],[],[],[],[16 80 20 0.25 0.25 0.5 0.5 -5 10],...
13  [24 160 60 1 1 2 2 5 20],@suhpaconstraints,options)
```

Ferguson spline wing:

```
Hermitevars9 =

   22.5912  107.7776   30.7826   0.6008   0.2776   0.9463
    1.0031    2.9569   12.6210

Hermitedrag9 =

   21.4888
```

The aerofoils and pressure profiles are shown in Figure 12.6. The drag achieved is in fact higher than the NACA-based design due to the Ferguson parameterization producing stronger adverse pressure gradients. There is more flexibility in the parameterization, but in this case it is not as appropriate for the design problem.

So far we have used the same aerofoil along the entire span (with angle-of-attack adjustment and linear scaling), but we can optimize the aerofoil at each of the four wing section joints. The search space is now very large, with 6×4 Ferguson aerofoil variables plus b, t_R and t_T. We therefore limit ourselves to a local search with the previous constant section design as a start point. The code for this search is shown in Listing 12.3 and yields a drag of 20.7576 N. The resulting design variables are shown in Table 12.3, and the aerofoils, with pressure contours, in Figure 12.7. Adding further flexibility to our parameterization by allowing the aerofoil shape to change along the span has improved the drag. However, the NACA four-digit parameterization is still clearly better for this design. With 27 variables we cannot be sure that there is not a better design *somewhere* in this vast design space. This is why more parsimonious parameterizations are often better; the design space is easier to search.

12.5   Improving the Design

It could be considered something of a failure not to improve on a design formulation from the 1930s, and, indeed, we can do better, by producing a parameterization designed specifically for

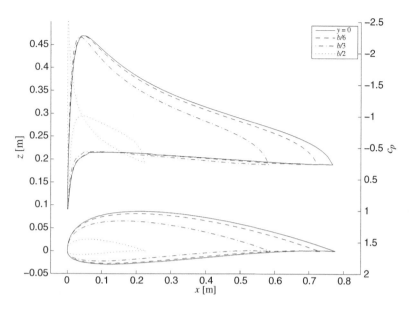

Figure 12.6 Aerofoils and corresponding pressure profiles after the nine-variable Ferguson spline-based optimization.

the problem at hand. 'Musculair 2' (Figures 12.8 and 12.9), designed and built by the German genius Gunther Rochelt in 1985, is the current holder of the human-powered aircraft world speed record. He used the Wortmann FX 76-MP-120 aerofoil for his wing design. Clearly, this is likely to be a good starting point. We, in the current SUHPA design, therefore adopted the strategy of fitting a parametric NURBS aerofoil to the FX 76-MP-120 and used it as a starting

Listing 12.3 Sample MATLAB® code to optimize SUHPA's wing by varying the thickness, span and the Ferguson spline aerofoil definition at four spanwise locations.

```
 1  % start from nine-variable optimization
 2  start=[Hermitevars9 Hermitevars9(4:end) Hermitevars9(4:end)...
 3  Hermitevars9(4:end)];
 4  % upper and lower bounds
 5  lb=[16 80 20 0.25 0.25 0.5 0.5 -5 10...
 6    0.25 0.25 0.5 0.5 -5 10 0.25 0.25 0.5 0.5 -5 10...
 7    0.25 0.25 0.5 0.5 -5 10];
 8  ub=[24 160 60 1 1 2 2 5 20 1 1 2 2 5 20...
 9    1 1 2 2 5 20 1 1 2 2 5 20];
10  % set finite differnce step size
11  options=optimset('Display','iter','FinDiffRelStep',0.001,...
12  'TypicalX',start);
13  % call optimizer
14  [Hermitevars27,Hermitedrag27]=fmincon(@suhpadrag,start,[],[],[],[],...
15  lb,ub,@suhpaconstraints,options)
```

Table 12.3   The 27-variable optimized SUHPA wing design.

Parameter	Root	$b/6$	$b/3$	$b/2$
b (m)			22.92	
t_R (mm)			108.2	
t_T (mm)			30.2	
T_A^{upper}	0.678	0.563	0.451	0.380
T_A^{lower}	0.254	0.268	0.266	0.556
T_B^{upper}	0.797	1.300	0.807	0.866
T_B^{lower}	1.428	1.003	1.567	0.614
α_c (deg)	2.927	2.988	3.011	2.857
α_b (deg)	12.53	12.73	12.53	13.02

point for the design. The NURBS control points are shown in Figure 12.10. The z-locations and weights were optimized to minimize the error between the NURBS aerofoil and the FX 76-MP-120. These weights were then varied to produce an improved design for SUHPA, based on the aero-structural optimization process described in the previous sections. Comparing the optimized design and FX 76-MP-120 (and their pressure profiles) in Figure 12.10, we see how a long, weakly adverse pressure gradient has been produced on the upper surface of the SUHPA aerofoil, attempting to delay transition for as long as possible. A great deal of care needs to be taken over the surface finish of such a wing if the predicted laminar flow region is to

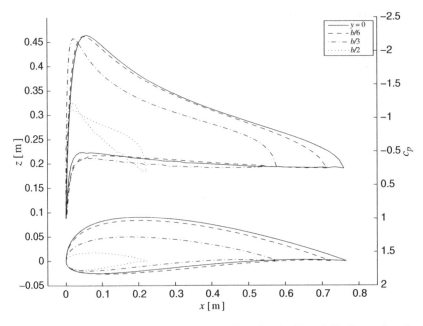

Figure 12.7   Aerofoils and corresponding pressure profiles after the 27-variable Ferguson spline-based optimization.

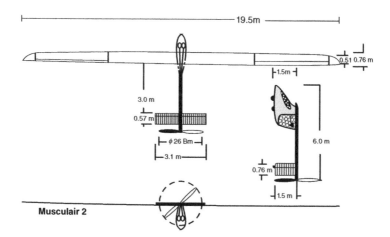

Figure 12.8 Musculair 2, designed by Gunther Rochelt.

Figure 12.9 Musculair 2 in flight (photograph by Jennifer Forrester).

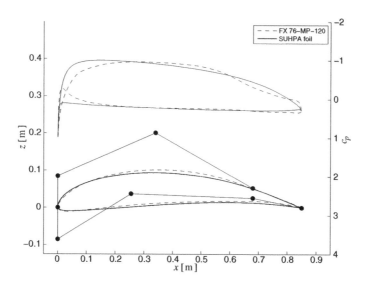

Figure 12.10 An aerofoil from SUHPA, showing the NURBS control points. Also plotted is the Wortmann FX 76-MP-120, upon which the NURBS parameterization was defined, and the pressure profiles for both aerofoils.

Figure 12.11 An image of SUHPA, coloured by CFD-simulated static pressure, with propeller wake streamlines. The image was produced using ANSYS Fluent, with thanks to Reece Gaj (University of Southampton).

be obtained in practice. This is an example of an optimizer exploiting the set-up of a problem, potentially finding loopholes, to the detriment of the performance of the final design. We have found a design that should perform well in perfect conditions, but is not robust to defects in, for example, surface roughness. The parsimonious NACA and (to a lesser extent) Ferguson parameterizations guard against this, but a flexible NURBS definition is more risky. As this is an experimental aircraft, designed with the aim of maximum speed, this risk is acceptable and necessary; however, varying operating conditions and in-service/manufacturing defects should normally be accounted for.

The wing optimization in this section has been based on simple 2D analysis codes (suitable for this high aspect ratio, low-cost wing, and useful as a re-creatable example), and so has not made use of any complex 3D geometry construction. However, we have shown in Chapter 9 how to create OpenNURBS/Rhino-Python models of parametric wings, via which the processes demonstrated in this chapter could take advantage of more complex 3D CFD solvers. We have employed two simple parameterizations of the wing; but, clearly, further complexity can be added using the variety of wing parameterizations covered in this text. Design sensitivities might also be included using algorithmic differentiation (Chapter 10). A 3D CFD simulation result for the whole SUHPA configuration is shown in Figure 12.11. The thousands of drag simulations conducted in the optimization procedures in this chapter would not be possible with expensive 3D analysis, and in such cases the designer needs to limit the number of design variables (even more so) and/or use surrogate models of the analyses (e.g. see Forrester *et al.* (2008)).

13

Epilogue: Challenging Topological Prejudice

There is an obvious hierarchy in terms of the construction of the external geometry of an airframe, and we have organized this text along those lines too.

At the lowest level one might find the geometrical primitives, the simple, generic curves and the surfaces that make up any parametric geometry. Next, we could imagine a tier comprising aircraft-specific primitives and their parameterizations – aerofoils, fuselage sections, planform geometries, and so on. At the next level we have major, 3D subcomponents, such as lifting surfaces and enclosure-type objects (fuselages, fairings, nacelles).

Then, a final level: the layout – or topology – of the overall airframe. But this text stops short of this top level, as does most of the rest of the literature and industrial practice. This is a key area of research, in its infancy today.

The design of passenger airliners is the most compelling illustration of this: topologically speaking, there is no difference between the airframes of the latest generation of airliners and, say, the Boeing 707. In fact, one might have to go back as far as the biplane airliners of the inter-war era to see significant differences in layout.

The main reason for this is to do with risk in an age of multibillion dollar major development programmes occurring once every decade or so. The big airframe integrators essentially bet the company on each completely new airliner design, and the resulting risk aversion understandably makes their approach conservative.

But is there solid scientific evidence that the conventional airframe layout (that which everyone would produce when asked to 'draw an aeroplane') is indeed the globally optimal solution? Designers of research aircraft and unmanned platforms, burdened by many fewer constraints than the airliner engineers, seem to disagree – Burt Rutan's Proteus, shown in Figure 13.1, is an example.

The answer to the question above is, 'almost certainly not'. However, at the moment we do not have the geometrical tools (at least not in a 'ready for the prime-time' state) to produce a *fully parametric* full airframe model capable of covering the layout search space. This would have to be a geometry model that is capable of producing a 'tube-and-wing'-type aeroplane in

Aircraft Aerodynamic Design: Geometry and Optimization, First Edition. András Sóbester and Alexander I J Forrester.
© 2015 John Wiley & Sons, Ltd. Published 2015 by John Wiley & Sons, Ltd.

Figure 13.1   Scaled composites Proteus (NASA image).

response to one set of inputs, a Proteus (Figure 13.1) to another, and completely unforeseen, weird and wonderful geometries to others. Moreover, the resulting design space should, ideally, have some form of continuity, in order to enable efficient optimization.

Geometry modelling, however, is not the only technology falling short, at the moment. Only a very sophisticated flow simulation, an automated and detailed structural and systems design process, coupled with a detailed life cycle cost model could guide an optimization process between the geometries produced by a topologically parametric model, and we are a long way away from such a capability.

Returning to parametric geometry modelling, however, here is then the challenge of the coming decades: a concise, robust and flexible mathematical formulation that will allow us to either conclude that the designers of, say, the Douglas DC-2 found *the* optimum layout geometry or to enable the automatic generation of better, bespoke topologies.

Two words are very important here. The first is *automatic*. We are looking for a design process free of topological prejudice, one that comes without any built-in bias towards one solution or another. The second is *bespoke*. At present, individual elements of an aircraft design are customized for a given mission – the entire system (including the topology) is not. In a world heading toward new materials, distributed (probably electrical) propulsion paradigms and strict environmental impact constraints, this is increasingly becoming a missed opportunity.

References

Abbott, I. H. and von Doenhoff, A. E. (1959) *Theory of Wing Sections*, Dover Books on Aeronautical Engineering, Dover Publications.

Anderson, J. D. (2005) *Introduction to Flight*, McGraw-Hill, 5th edition.

Barbarino, S., Bilgen, O., Ajaj, R. M. *et al.* (2011) A review of morphing aircraft. *Journal of Intelligent Material Systems and Structures*, **22** (9), 823–877.

Bauer, J. E., Clarke, R. and Burken, J. J. (1995) Flight test of the X-29A at high angle of attack: flight dynamics and controls, Technical Paper 3537, NASA.

Broyden, C. G. (1970) The convergence of a class of double-rank minimization algorithms. *Journal of the Institute of Mathematics and its Applications*, **6**, 76–90.

Buckley, H. P., Beckett, Y. Z. and Zingg, D. W. (2010) Airfoil optimization using practical aerodynamic design requirements. *Journal of Aircraft*, **47** (5), 1707–1719.

Cliff, S. E., Reuther, J. J., Saunders, D. A. and Hicks, R. M. (2001) Single-point and multipoint aerodynamic shape optimization of high-speed civil transport. *Journal of Aircraft*, **38** (6), 997–1005.

Coquillart, S. (1990) Extended free-form deformation: a sculpturing tool for 3D geometric modeling. *ACM SIGGRAPH Computer Graphics*, **24** (4), 187–196.

Demasi, L., Dipace, A., Monegato, G. and Rauno, C. (2014) An invariant formulation for the minimum induced drag conditions of non-planar wing systems. *AIAA 2014-0901*, 1–33, doi:10.2514/6.2014-0901.

Drela, M. (1989) XFOIL: an analysis and design system for low Reynolds number airfoils. *Conference on Low Reynolds Number Airfoil Aerodynamics, University of Notre Dame.*

Drela, M. (2014) *Flight Vehicle Aerodynamics*, MIT Press, London.

Drela, M. and Giles, M. (1987) Viscous–inviscid analysis of transonic and low Reynolds number airfoils. *AIAA Journal*, **25** (10), 1347–1355.

Drela, M. and Youngren, H. (2011) XFOIL 6.9 user primer, URL http://web.mit.edu/drela/Public/web/xfoil/.

Farin, G. (1999) *NURBS: From Projective Geometry to Practical Use*, Aerospace Science Series, AK Peters, Natick, MA, 2nd edition.

Ferguson, J. (1964) Multivariable curve interpolation. *Journal of the Association for Computing Machinery*, **11** (2), 221–228.

Feynman, R., Leighton, R. B. and Sands, M. (1964) *Mainly Electromagnetism and Matter*, volume II of *The Feynman Lectures on Physics*, Addison-Wesley, Reading, MA.

Forrester, A. I. and Keane, A. J. (2009) Recent advances in surrogate-based optimization. *Progress in Aerospace Sciences*, **45** (1), 50–79.

Forrester, A. I. J., Sóbester, A. and Keane, A. J. (2008) *Engineering Design via Surrogate Modelling*, John Wiley & Sons.

Forth, S. (2006) An efficient overloaded implementation of forward mode automatic differentiation in MATLAB. *ACM Transactions on Mathematical Software*, **32** (2), 195–222.

Gielis, J. (2003) A generic geometric transformation that unifies a wide range of natural and abstract shapes. *American Journal of Botany*, **90** (3), 333–338.

Griewank, A. (2000) *Evaluating Derivatives: Principles and Techniques of Algorithmic Differentiation*, Frontiers in Applied Mathematics, SIAM, Philadelphia, PA.

Harris, C. D. (1990) NASA supercritical airfoils – a matrix of family-related airfoils, Technical Paper 2969, NASA.

Hicks, R. M. and Henne, P. A. (1978) Wing design by numerical optimization. *Journal of Aircraft*, **15**, 407–412.

Hooke, R. and Jeeves, T. A. (1961) "Direct search" solution of numerical and statistical problems. *Journal of the ACM*, **8** (2), 212–229.

Houghton, E. L. and Carpenter, P. W. (1994) *Aerodynamics for Engineering Students*, Edward Arnold, London, 2nd edition.

Hsu, W. M., Hughes, J. F. and Kaufman, H. (1992) Direct manipulation of free-form deformations. *ACM SIGGRAPH Computer Graphics*, **26** (2), 177–184.

Hu, H., Tamai, M. and Murphy, J. (2008) Flexible-membrane airfoils at low Reynolds numbers. *Journal of Aircraft*, **45** (5), 1767–1778.

Isikveren, A. T. (2002) Quasi-analytical modelling and optimisation techniques for transport aircraft design, Ph.D. thesis, Royal Institute of Technology (KTH), Stockholm, Sweden.

Jacobs, E. N. and Pinkerton, R. M. (1935) Tests in the variable-density wind tunnel of related airfoils having the maximum camber unusually far forward, Report No. 537, NACA.

Jacobs, E. N., Ward, K. E. and Pinkerton, R. M. (1933) The characteristics of 78 related airfoil sections from tests in the variable-density wind tunnel, Report No. 460, NACA.

Jameson, A. (1999) Re-engineering the design process through computation. *Journal of Aircraft*, **36** (1), 36–50.

Jasak, H., Jemcov, A. and Tukovic, Z. (2007) OpenFOAM: a C++ library for complex physics simulations, in *International Workshop on Coupled Methods in Numerical Dynamics*, pp. 1–20.

Johnsen, F. A. (2013) *Sweeping Forward – Developing & Flight Testing the Grumman X-29 Forward Swept Wing Research Aircraft*, NASA Aeronautics Book Series, NASA.

Jones, D. R., Schonlau, M. and Welch, W. J. (1998) Efficient global optimization of expensive black-box functions. *Journal of Global Optimization*, **13** (4), 455–492.

Juhász, I. and Hoffmann, M. (2004) Constrained shape modification of cubic B-spline curves by means of knots. *Computer-Aided Design*, **36** (5), 437–445, doi:http://dx.doi.org/10.1016/S0010-4485(03)00116-7, URL http://www.sciencedirect.com/science/article/pii/S0010448503001167.

Kulfan, B. (2007) Aerodynamics of sonic flight, Seminar, Tohoku University.

Kulfan, B. M. (2006) "Fundamental" parametric geometry representations for aircraft component shapes, in *AIAA 2006-6948*, pp. 1–45.

Kulfan, B. M. (2008) Universal parametric geometry representation method. *Journal of Aircraft*, **45** (1), 142–158.

Kulfan, B. M. (2010) Modification of CST airfoil representation methodology, unpublished note, http://brendakulfan.com/docs/CST8.pdf. Accessed: 23 April 2013.

Kulfan, B. M. and Bussoletti, J. E. (2006) "Fundamental" parametric geometry representations for aircraft component shapes, in *AIAA 2006-6948*, pp. 1–45.

Kulfan, R. M. (1973) High transonic speed transport aircraft study summary report, Report CR-2465, NASA.

Ladson, C. L., Brooks, C. W., Hill, A. S. and Sproles, D. W. (1996) Computer program to obtain ordinates for NACA airfoils, Technical Memorandum 4741, NASA.

Lissaman, P. (1983) Low-Reynolds-number airfoils. *Annual Review of Fluid Mechanics*, **15**, 223–239.

Loftin, L. K. (1948) Theoretical and experimental data for a number of NACA 6A-series airfoil sections, Report No. 903, NACA.

Mason, W. (2009) Lecture notes, Virginia Tech University.

McKay, M., Conover, W. and Beckman, R. (1979) A comparison of three methods for selecting values of input variables in the analysis of output from a computer code. *Technometrics*, **21**, 239–245.

McLean, D. (2012) *Understanding Aerodynamics: Arguing from the Real Physics*, volume 69 of *Aerospace Series*, John Wiley & Sons.

Meredith, P. T. (1993) Viscous phenomena affecting high-lift systems and suggestions for future CFD development, in *AGARD Conference Proceedings*, volume 515, pp. 19–1–19–8.

Nelder, J. A. and Mead, R. (1965) A simplex method for function minimization. *The Computer Journal*, **7** (4), 308–313.

Ning, A. and Kroo, I. (2010) Multidisciplinary considerations in the design of wings and wing tip devices. *Journal of Aircraft*, **47** (2), 534–543.

Null, W. and Shkarayev, S. (2005) Effect of camber on the aerodynamics of adaptive-wing micro air vehicles. *Journal of Aircraft*, **42** (6), 1537–1542.

Nurdin, A., Bressloff, N. W. and Keane, A. J. (2012) Shape optimisation using CAD linked free-form deformation. *The Aeronautical Journal*, **116** (1183), 915–939.

Patterson, E. W. and Braslow, A. L. (1958) Ordinates and theoretical pressure-distribution data for NACA 6- and 6A-series airfoil sections with thicknesses from 2 to 21 and from 2 to 15 percent chord, respectively, Report TN-4322, NACA.

Phillips, W. F. (2010) *Mechanics of Flight*, John Wiley & Sons.

Piegl, L. A. and Tiller, W. (1996) *The NURBS Book*, Monographs in Visual Communication, Springer.

Powell, S., Sóbester, A. and Joseph, P. (2012) Fan broadband noise shielding for over-wing engines. *Journal of Sound and Vibration*, **331** (23), 5054–5068.

Powell, S. R. (2012) Applications and enhancements of aircraft design optimization techniques, Ph.D. thesis, University of Southampton, Faculty of Engineering and the Environment.

Raymer, D. P. (2006) *Aircraft Design: A Conceptual Approach*, AIAA.

Robinson, G. M. and Keane, A. J. (2001) Concise orthogonal representation of supercritical airfoils. *Journal of Aircraft*, **38**, 580–583.

Rudolph, P. K. C. (1996) High-lift systems on commercial subsonic airliners, NASA Contractor Report 4746.

Salomon, D. (2006) *Curves and Surfaces for Computer Graphics*, Springer, New York, NY.

Sederberg, T. W. and Parry, S. R. (1986) Free-form deformation of solid geometric models. *ACM SIGGRAPH Computer Graphics*, **20** (4), 151–160.

Selig, M. (2003) Low Reynolds number airfoil design, in *VKI Lecture Series – Low Reynolds Number Aerodynamics on Aircraft Including Applications in Emerging UAV Technology*, von Karman Institute for Fluid Dynamics, von Karman Institute for Fluid Dynamics (VKI) Lecture Series.

Sóbester, A. (2009) Concise airfoil representation via case-based knowledge capture. *AIAA Journal*, **47** (5), 1209–1218, URL http://eprints.soton.ac.uk/66386/.

Sóbester, A. and Keane, A. J. (2007) Airfoil design via cubic splines – Ferguson's curves revisited, in *AIAA Infotech@Aerospace 2007 Conference and Exhibit*, Rohnert Park, CA.

Sóbester, A. and Powell, S. (2013) Design space dimensionality reduction through physics-based geometry reparameterization. *Optimization and Engineering*, **14** (1), 37–59, URL http://eprints.soton.ac.uk/210952/.

Sobieczky, H. (1998) *Parametric Airfoils and Wings*, volume 68 of *Notes on Numerical Fluid Mechanics*, Vieweg Verlag, Wiesbaden.

Sobieczky, H., Dougherty, F. C. and Jones, K. (1990) Hypersonic waverider design from given shock waves, in *Proceedings of the First International Waverider Symposium*, University of Maryland.

Straathof, M. (2012) Shape parameterization in aircraft design: a novel method, based on B-splines, Ph.D. thesis, TU Delft, Faculty of Aerospace Engineering.

Taylor, G. I. and Maccoll, J. W. (1933) The air pressure on a cone moving at high speeds. I. *Proceedings of the Royal Society of London, Series A*, **139** (838), 278–297.

Thomas, I. (1939) *Selections Illustrating the History of Greek Mathematics (with English translation)*, Harvard University Press, Cambridge, MA.

Trimble, S. (2013) Extra design changes hike type's empty weight to 220t. *Flight International*, **182** (5378), 13.

Ursache, N. M., Keane, A. J. and Bressloff, N. W. (2006) Design of postbuckled spinal structures for airfoil camber and shape control. *AIAA Journal*, **44** (12), 3115–3124.

van Dam, C. (2002) The aerodynamic design of multi-element high-lift systems for transport airplanes. *Progress in Aerospace Sciences*, **38** (2), 101–144.

Walter, S. F. and Lehmann, L. (2013) Algorithmic differentiation in Python with AlgoPy. *Journal of Computational Science*, **4** (5), 334–344.

Wauquiez, C. (2000) *Shape Optimization of Low Speed Airfoils using Matlab and Automatic Differentiation*, Licentiates Thesis, Royal Institute of Technology, Stockholm.

Whitcomb, R. T. (1974) Review of NASA supercritical airfoils, in *The Ninth Congress of the International Council of the Aeronautical Sciences, ICAS 74-10*.

Whitcomb, R. T. and Clark, L. R. (1965) An airfoil shape for efficient flight at supercritical Mach numbers, Technical Report TM X-1109, NASA.

Zhu, C., Byrd, R. H. and Nocedal, J. (1997) L-BFGS-B: Algorithm 778: L-BFGS-B, FORTRAN routines for large scale bound constrained optimization. *ACM Transactions on Mathematical Software*, **23** (4), 550–560.

Index

Pages numbers in **bold** refer to figures, and `Courier` to computer code. Entries in `Courier` refer to MATLAB® or Python functions.

Aircraft Aerodynamic Design: Geometry and Optimization, First Edition. András Sóbester and Alexander I J Forrester.
© 2015 John Wiley & Sons, Ltd. Published 2015 by John Wiley & Sons, Ltd.

Printed and bound by CPI Group (UK) Ltd, Croydon, CR0 4YY

12/01/2025

14624500-0001